CHASSIS ENGINEERING

BY HERB ADAMS

HPBooks

HPBOOKS
Published by the Penguin Group
Penguin Group (USA) Inc.
375 Hudson Street, New York, New York 10014, USA
Penguin Group (Canada), 90 Eglinton Avenue East, Suite 700, Toronto, Ontario M4P 2Y3, Canada
(a division of Pearson Penguin Canada Inc.)
Penguin Books Ltd., 80 Strand, London WC2R 0RL, England
Penguin Group Ireland, 25 St. Stephen's Green, Dublin 2, Ireland (a division of Penguin Books Ltd.)
Penguin Group (Australia), 250 Camberwell Road, Camberwell, Victoria 3124, Australia
(a division of Pearson Australia Group Pty. Ltd.)
Penguin Books India Pvt. Ltd., 11 Community Centre, Panchsheel Park, New Delhi—110 017, India
Penguin Group (NZ), 67 Apollo Drive, Rosedale, North Shore 0632, New Zealand
(a division of Pearson New Zealand Ltd.)
Penguin Books (South Africa) (Pty.) Ltd., 24 Sturdee Avenue, Rosebank, Johannesburg 2196,
South Africa

Penguin Books Ltd., Registered Offices: 80 Strand, London WC2R 0RL, England

Cover design by Beth Bender
Interior photos and illustrations by Herb Adams and Michael Lutfy

First edition: November 1992

Library of Congress Cataloging-in-Publication Data

Adams, Herb.
Chassis engineering : chassis design building & tuning
for high performance handling / by Herb Adams
p. cm.
Includes index.
ISBN 978-1-55788-055-0
1. Automobiles—Chassis. 2. Automobiles—Performance.
I. Title.
TL255.A23 1993
629.24—dc20 92-37394
CIP

PRINTED IN THE UNITED STATES OF AMERICA

35 34 33 32 31 30 29

ABOUT THE AUTHOR

Herb Adams began his automotive engineering career when he joined the engineering staff at Pontiac after graduating from the General Motors Institute in 1957. At Pontiac, he designed and developed a variety of experimental and production engines, most notably the Pontiac Trans Am. In 1973, he left General Motors to begin his own automotive engineering and consulting business, while building his own racing cars which he drove in several professional racing series, including the SCCA Trans-Am. His company, VSE, specializes in handling and appearance items for American sports cars, such as the Camaro, Firebird, Corvette and Mustang. He also serves as an engineering consultant to Goodyear, Pontiac, Oldsmobile, Koni, Appliance Wheels and Cincinnati Microwave. In recent years, he has written numerous articles on chassis design and building for *Circle Track* magazine. He lives in Northern California. ■

Contents

INTRODUCTION

The purpose of this book is to explain some of the chassis engineering aspects related to the design, build and testing of high performance automobiles. One area of race car performance that still seems difficult to explain is suspension and handling. Although many enthusiasts can tell you what changes will produce what results, they may not be able to tell you *why* these changes produce these results. In order to understand the whys of suspension and handling, you need to study the physics involved. Since not all enthusiasts have an engineering or physics background, this book will attempt to discuss suspensions and handling in terms the layperson can understand.

Good handling could be described as going around corners faster while improving driver control. Suspensions are involved in the handling equation because they control how the tires work against the track. Because the tires are the only link between the car and

the track, they are the key to improved handling. Almost all of the suspension variables are related to how well a car's tires react to the ground.

Just as in the case of engine tuning, where very small changes can have a dramatic effect on horsepower output, very slight changes in the suspension of your car can have significant effects on your handling. Even a one-degree change in camber can have a measurable effect on performance, so a high degree of precision is required to set-up a suspension properly.

Another aspect of suspension tuning that should be stressed at the beginning is the concept of optimizing settings. For example, if you increased your tire pressure from 30 to 40 psi, and you measured some improvement, there is no reason to assume that increasing the pressure from 40 to 50 psi will have an equal improvement. If the tire operates best at 40 psi, increasing the pressure to 50 psi could actually reduce its performance. Every suspension adjustment and setting has an optimum value. We know that more valve timing in an engine can increase power up to a certain point and then further increases will lose power. The same concept is true in suspension tuning.

This is not an engineering book in the classical sense, but rather an explanation of how engineering principles are applied to the automobile chassis. This application of engineering to the automobile chassis is a very specialized field of study, so it is not widely practiced. However, the interest in chassis engineering is very broad, because there are so many automotive enthusiasts interested in improving the cornering performance of their cars. By studying the engineering relationships, and the examples given in this book, most enthusiasts with average mechanical skills should be able to make significant improvements in the cornering performance of their car, regardless of the type or application, be it street, circle track or road racing. Because of the many different types of vehicles and racing applications, it would be nearly impossible to do a true "how-to" book. The idea is to present the information in a universal manner, so it can be applied to suit any need.

The information contained in this book was accumulated over many years, the result of numerous contributions by top engineers and racers. It is not unusual for an engineer or a car builder to struggle for months to solve a problem. And, once they do find the answer, it is simple to follow their lead. During the past 30 years, there has been a tremendous increase in the available knowledge on car handling, and this book attempts to present some of this information. Therefore, when "we" is used throughout this book, it refers to all those people who have helped to accumulate this information.

To help explain the handling concepts, I have organized this book using the "building block" approach. The first chapters present the most basic concepts. The later chapters use these basic concepts to explain the more complicated relationships and solve common problems. Don't read the chapters out of order, because you need to fully understand tire characteristics before you can understand weight distribution dynamics, or you need to know how tire loadings, stabilizer bars and spring rates affect handling before you can correct any oversteer or understeer problems. The study of chassis engineering is complex and can be difficult to grasp. Hopefully, this book will make it easier.■

Herb Adams

TIRE CHARACTERISTICS 1

The tires on your car have more effect on its handling than any other component. Understanding how the tires work on your car is absolutely necessary in order to understand why a car handles the way it does, or more importantly, to change how it handles. To simplify the study of how tires work, we can limit the analysis to their *input-output characteristics*, since these are the most important factors that affect handling. This type of analysis is called the *black-box method* , because we do not concern ourselves with what is going on inside the "box," or tire in this case.

Vertical Load—The input for tire performance is the *vertical load*, or weight, on the tire. The dynamics of the car in motion cause this load to vary continuously. By tuning the chassis, it is possible to adjust how this vertical tire loading will change, and by knowing how the tire will respond to the change in loading, you will be able to predict the effect of the changes.

Traction—The output of a tire from a handling standpoint is its *traction* or how well it "sticks" to the ground. The traction between the tires and the ground determines how fast a car can accelerate, brake and/or corner.

TIRE TRACTION VS. LOAD

It is necessary to know how the tire translates the input into output to understand how a car will handle the way it does. In other words, you need to know how changes in vertical load (input) affect the traction (output). The relationship between tire input and output forces is different for every tire but, more importantly, the relationship changes dramatically as the vertical loading is changed. This changing relationship is the major reason why the study of handling is often confusing.

Although the relationship between vertical load and traction for any given tire is continually changing, the interaction between the two will follow a curve similar to that shown in Figure 1-1. It is not necessary to have the tire performance curve for your specific tires because we are looking for the tire characteristics, not the exact values. The shape of the tire performance curve is what is important.

Different tires will have performance curves with different shapes or values, but they all will have a curve that results in a smaller increase in traction as the vertical load is increased. We call this loss of relative traction a loss in the tire's *cornering efficiency*.

Good handling depends on optimizing how your car uses its tires. You need to know what your tires want if you expect to maximize their performance.

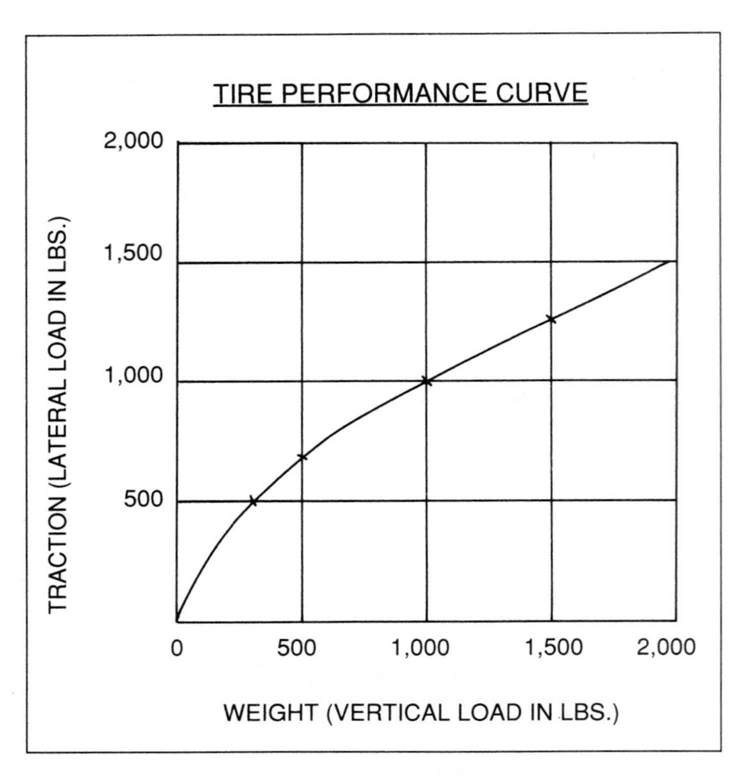

Figure 1-1. This is a tire performance curve. The amount of traction available from any given tire is dependent on how much weight is on the tire. As weight is increased, the traction also increases. The important thing that must be recognized however, is that the increase in traction becomes less and less as the weight is increased.

By making a chart of the tire performance curve (Chart 1-1) it is possible to see how the cornering efficiency of the tire decreases as the vertical loading is increased. The vertical load and traction readings were taken from the tire performance curve (Figure 1-1). Efficiency is output (traction) divided by input (vertical load) as shown on Chart 1-1.

By examining Chart 1-1, you can see that with a cornering efficiency of 140%, it would be possible for a car to corner at 1.40 *g*'s. When the cornering efficiency is only 75%, the same car would corner at .75 *g*'s.

By looking at tire cornering efficiency, you can easily see that you will get the most cornering power as a percentage of the vertical load when the vertical loads are lower. A tire's cornering efficiency reduces quickly when it is asked to support more and more weight. This characteristic of any tire is a key element in understanding why cars handle the way they do. Understanding this aspect of a tire's performance is necessary to analyze the more complex conditions that a car experiences during actual driving conditions.

Chart 1-1

VERTICAL LOAD VS. CORNERING EFFICIENCY

Vertical Load	Traction Available	Traction / Vertical Load	Factor	Cornering Efficiency
500	700	$\frac{700}{500}$	1.40	140%
1000	1000	$\frac{1000}{1000}$	1.00	100%
1500	1250	$\frac{1250}{1500}$	.83	83%
2000	1500	$\frac{1500}{2000}$	.75	75%

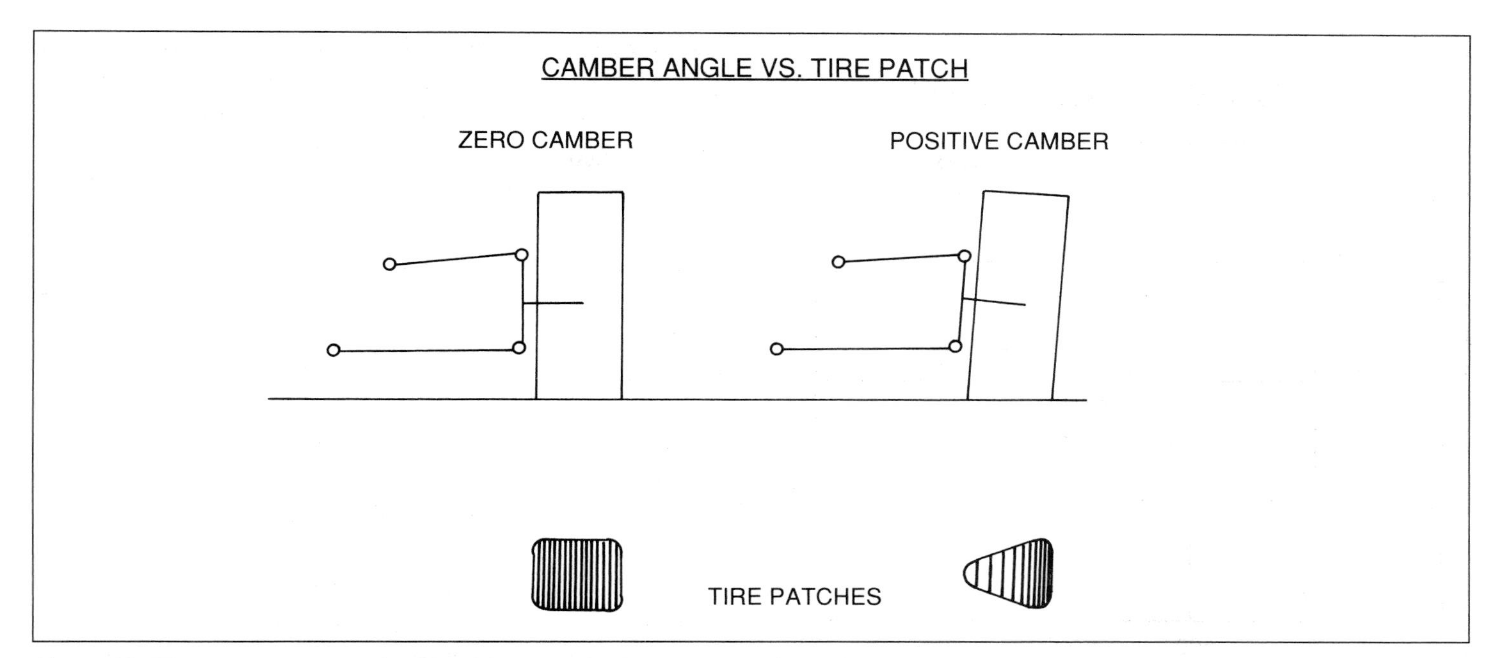

Figure 1-2. When a tire is perpendicular to the ground, it has zero camber angle. This condition provides the biggest tire patch for the most possible traction. Equally important is that the unit-loading on each part of the tire patch is more evenly distributed when the tire has zero camber angle. As shown here, a positive camber angle will result in less of a tire patch.

Tire Factors—When analyzing your car's handling, tire factors such as contact patch, tread depth, aspect ratio, etc., must be considered because they change how much traction your tires can provide for a given vertical load. These factors raise or lower the traction curve and they can cause the shape of the curve to change. When the same size and type of tire is used at all four locations, these factors affect how much traction is available.

CAMBER ANGLE & CONTACT PATCH

A tire will provide the maximum traction at any given vertical load when it is perpendicular to the ground. This is called *zero camber angle* (Figure 1-2). When the tire is perpendicular to the ground, its *contact patch* is bigger than when it is at any other angle. Contact patch is the area of the tire in direct contact with the road surface. If a tire is tilted out at the top, it has *positive camber.* This condition reduces the tire contact patch and the tire will not provide as much traction as when it is perpendicular. *Negative camber,* when the top of the tire is tilted inward, is often dialed in to compensate for the moving or bending (known as deflection) of suspension parts. When it is used, the result is to have zero camber angle when maximum tire traction is needed.

CIRCLE OF TRACTION

The Circle of Traction concept is based on the fact that a tire has only a certain amount of traction at any given time. This total amount of traction is dependent on the weight on the tire, the track conditions, the weather, etc. When studying the Circle of Traction, the total amount of traction is considered constant. What the Circle of Traction shows is how this total amount of traction is distributed between cornering forces and acceleration or braking forces. If you only have so much traction available, deciding how to use this traction can have an important effect on how well a car handles. The Circle of Traction concept says that the amount of cornering force available for a tire will be reduced by whatever amount of the total traction is also used for acceleration or braking.

How it Works—If you could view the tire contact patch as it moves along the roadway, you could see how this Circle of Traction operates. The total traction capability can be represented by an arrow on a circular graph. This arrow represents the available traction and it can be pointed in any direction (Figure 1-3).

For example, if a sample tire had 1000 lbs. of load on it, and its cornering efficiency was 100%, it would provide 1000 lbs. of traction. This 1000 lbs. of traction is available in any direction–pure cornering, acceleration or braking. But unfortunately, it is not

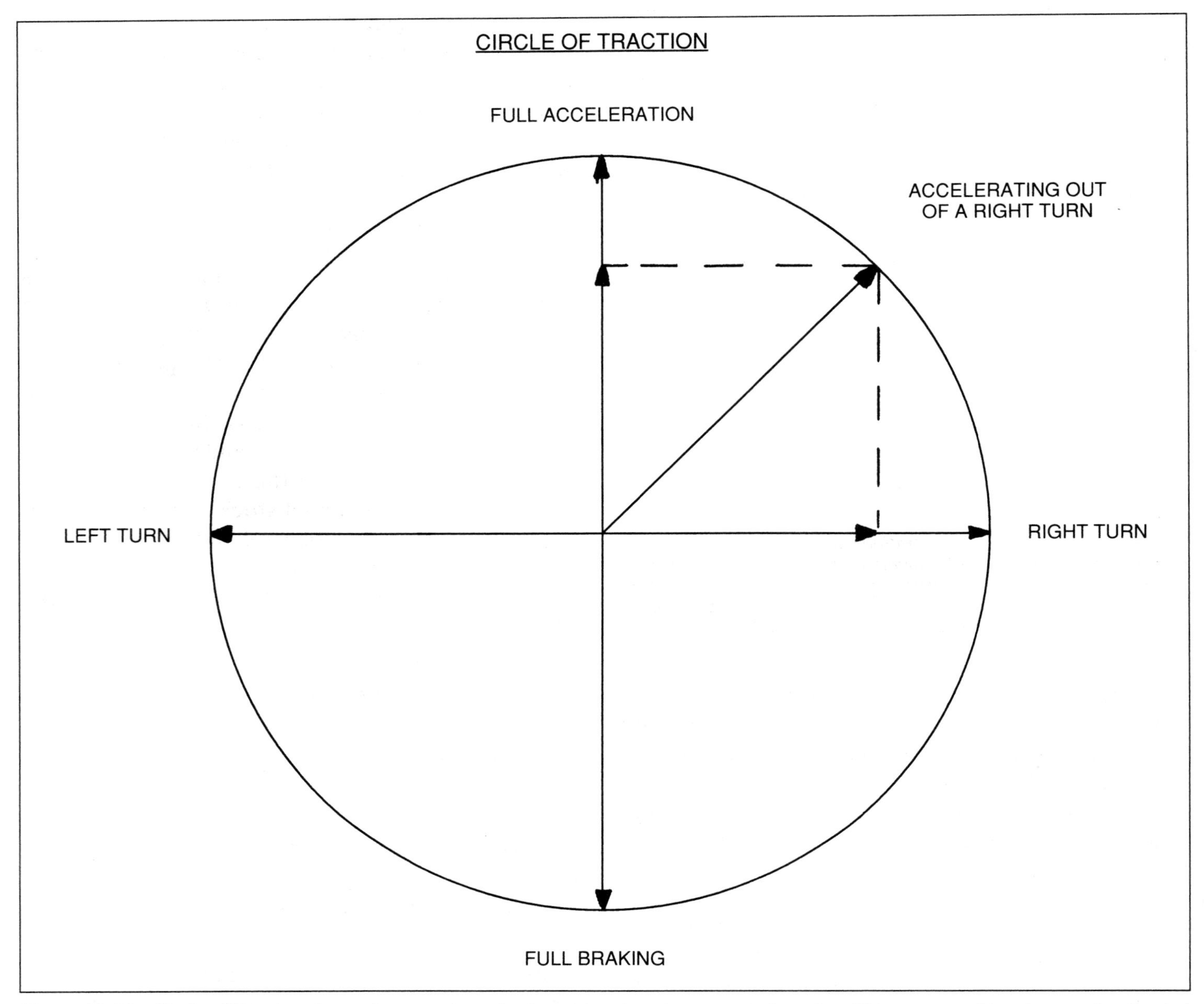

Figure 1-3. The Circle of Traction shows that any given tire has only a certain amount of traction. This amount of traction can be directed in any direction, but if any of it is used for acceleration, less will be available for cornering.

available in any two directions at once at its full traction of 1000 lbs. If some of the total 1000 lbs. is used in acceleration, less than 1000 lbs. will be available for cornering. The total is not additive, but is a vector amount that can be used in combination as shown on the Circle.

Every driver has experienced the effects on handling of this condition. When exiting a turn, a car that has normal understeer will have oversteer at full throttle. The reason for this change in cornering attitude on the same car in the same corner can be explained by looking at the Circle of Traction. As the driver asks the rear tires to absorb more acceleration force, there is less cornering force available at the rear, so the car has more oversteer as the driver applies more power.

Acceleration Effects—The extreme example of this condition is a car making a wheel-spinning start. If there is enough power to cause both rear wheels to break traction, all of the tires' traction is being used in the accelerating direction. As shown in the Circle of Traction diagram (Figure 1-3), this condition results in zero cornering power available from the tires to restrain the car from side loadings. The results of this lack of lateral force from the tires will cause the rear of the car to "fish-tail."

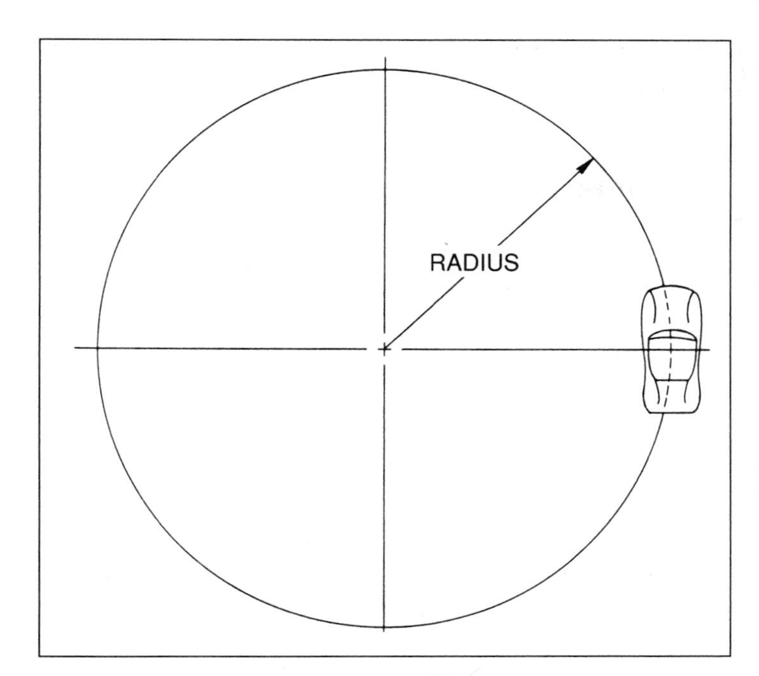

Figure 1-4. A car's lateral acceleration can be measured on a skidpad, which is a flat area of pavement usually 200 to 300 feet in diameter. The car is driven around the circle as fast as possible, the time is measured, and the lateral acceleration, expressed in g's, is calculated from the time and size of the circle.

Braking Effects—The effects of braking are similar but opposite. We know that locking the front tires will make the front-end go straight regardless of the steering angle of the front wheels. When the front tires are locked up, all of their available traction is being used to absorb the braking forces, so there is none left to provide the cornering power needed to make the car turn.

On a moving car, the distribution of acceleration, cornering and braking forces is constantly changing. If the driver and the chassis tuner are aware of how these changes affect the balance of the car, they will better understand what is needed to tune the chassis for maximum performance under all of these driving conditions.

g-FORCES

Tire and handling performance is described in terms of *g*-force. One *g* is simply the force equal to gravity here on Earth. If an object is said to weigh 100 pounds, the force of gravity on it equals 100 pounds. If this object is subjected to a second force of 80 pounds, we would say it has an .8 *g*-force acting on it. This system of describing forces affecting common objects—like automobiles—is more convenient than using pounds of force, since it eliminates the need to recognize the weight of the object. For instance, a 3000-lb. cornering force acting on a 3000-lb. car would be a 1.0 *g* load. The same 3000-lb. force on a 4000-lb. car would be a .75 *g* load. By describing cornering forces in *g*'s, various cars can be compared equally, regardless of their individual weights.

SKIDPAD TESTING

Many magazine road tests include a measure of a car's cornering power, or *lateral acceleration*. It is measured on a skidpad and expressed in *g*'s. A skidpad is a flat area of pavement with a painted circle, usually 200 to 300 feet in diameter. The car is driven around the circle as fast as possible without spinning out, the time is measured, and the lateral acceleration is calculated from the time and size of the circle. A typical late-model stock Corvette can corner at .84 *g*'s, a very respectable figure. A road-racing sedan, however, does considerably better. Our Trans-Am race car produced 1.15 *g* in skidpad testing.

A simplified formula for determining a car's cornering power on a skidpad is:

$$g = \frac{1.225 \times R}{T^2}$$

R = Radius of the turn in feet

T = Time in seconds required to complete a 360-degree turn

Plugging in some "real world" numbers shows how the formula works. For example, if a car takes 12 seconds per lap on a 100-foot radius skidpad, the computation is as follows:

$$g = \frac{1.225 \times 100}{12 \times 12}$$

$$g = \frac{122.5}{144}$$

$$g = .85$$

This means that the car is cornering at a force equal to 85/100 of the force of gravity. Now, let's take all of this information, and apply it to a discussion on weight distribution and dynamics. ■

WEIGHT DISTRIBUTION & DYNAMICS

In Chapter 1, we discussed how the amount of traction available from a tire is related to how much vertical load, or weight is on it. By using the tire performance curve (Figure 1-1) shown in Chapter 1, page 2, it is possible to determine how much traction is available at each individual tire. If you know how much vertical load there is, and how much traction is available, you can determine how much total cornering power the car has available. Knowing the individual tire traction limits can also tell you some of the handling characteristics such as *understeer* and *oversteer.*

Understeer & Oversteer—These are terms that describe how a car goes around a corner. If a car goes around a corner with the front of the car pointed toward the outside of the turn, it is said to understeer, or *push* through the turn. If a car goes around a corner with the rear of the car sliding toward the outside of the turn, it is said to oversteer and it is *loose*. For an illustrated example, see Figure 2-1.

WEIGHT DISTRIBUTION

A car's weight distribution is determined by how much weight is on each tire. These weights change due to *load transfer*. The changes in loading are the result of forces acting on the car. The following examples illustrate how some of these forces can change the individual vertical loads on each tire of a car.

Using this type of analysis is helpful in understanding how the static and dynamic weight distribution of a car can affect its handling characteristics. This example shows how the weights and therefore the traction available can change as a car

Weight distribution is how much weight, or load, each tire has on it at rest. However, when the car goes around a turn, weight will be transferred from the inside tires to the outside tires, which is known as lateral weight transfer. The more weight on a tire, the less traction it will have. With this knowledge, you can control the amount of weight transfer so your outside tires will have the maximum traction available. Understanding this relationship is a fundamental key to good handling. Photo by Michael Lutfy.

UNDERSTEER AND OVERSTEER

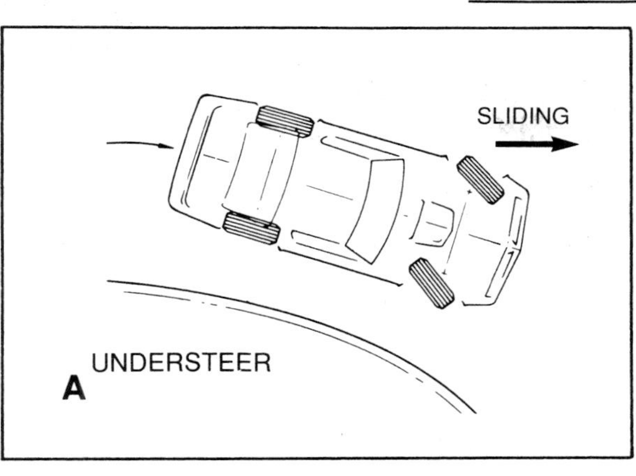

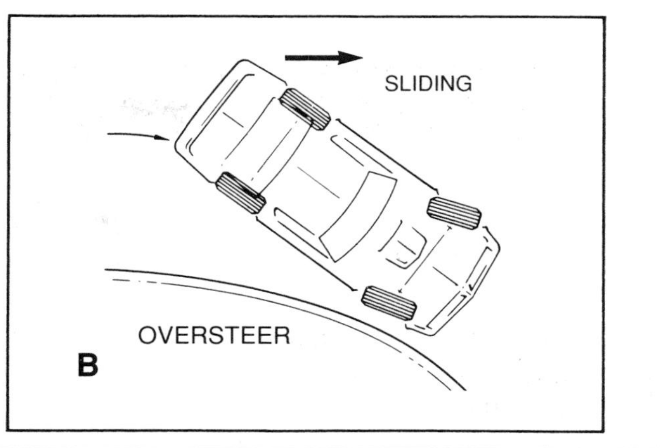

Figure 2-1. Understeer is the condition where a car needs more than normal front-wheel steering angle to go around a corner; the front-end of the vehicle tends to break loose and slide, or push toward the outside of a turn. Oversteer is when a car needs less than normal front-wheel steering angle to go around a corner; the rear end of the vehicle tends to break loose and slide outward.

moves. But the important concept to understand is that the traction available from a tire is dependent on its vertical load. The confusing aspect is that the percentage of traction improvement goes down as the load goes up. As we look at the following examples, keep in mind that the traction available values in the charts were taken from our tire performance curve, Figure 1-1, page 2.

Example One

Assume we have a car with the following weight distribution:

Car Weight:	**3000 lbs.**
Front End Weight:	**50%**
Left Side Weight Bias:	**0**
Load Transfer From Cornering:	**0**

Chart 2-1

Tire Location	Static Weight On Tire	Traction Available
Left Front	750 lbs.	850 lbs.
Right Front	750	850
Left Rear	750	850
Right Rear	750	850
Totals	3000 lbs.	3400 lbs.

This theoretical car has 750 lbs. on each wheel. Using the tire performance curve in Figure 1-1 on page 2, you can see that you would have 850 lbs. of traction available at each wheel. The total traction would be 3400 lbs. and the cornering force would be 1.13 *g*'s. See Chart 2-1. To find the total cornering force in *g*'s, you would divide the total traction by the total weight, from Chart 2-1, or:

$$\text{Total Cornering Force} = \frac{\text{Traction}}{\text{Weight}}$$

$$\text{Total Cornering Force} = \frac{3400}{3000} = 1.13\ g\text{'s}$$

Scales such as the ones pictured here can be used to determine the static weight on each tire of a car. Once you know the weight, you can take this information and compare it to the tire performance curve of your tires (such the one shown in Figure 1-1, p. 2) to determine the traction available. Of course, once the car is moving, these weights change continually and affect the traction available, so you need to understand the forces that control these changes. Photos by Michael Lutfy.

Chart 2-2				
Tire Location	**Static Weight On Tire**	**Lateral Weight Transfer**	**Weight On Tire During Cornering**	**Traction Available**
Left Front	750	-500	250	450
Right Front	750	+500	1250	1130
Left Rear	750	-500	250	450
Right Rear	750	+500	1250	1130
Totals	3000		3000	3160

This sounds pretty good until you realize that the weight will transfer from the inside tires to the outside tires as the car develops cornering force going around a corner.

Example Two

As soon as a car starts to go around a corner, its vertical tire loadings will change. Because of the cornering force, weight will be transferred from the inside tires to the outside tires (Figure 2-2). This change in loading is dependent on the cornering force (g's), the car track width (T), the height of the center of gravity (H) and the overall weight of the car (W). See Figure 2-3. Expressed as a formula:

$$\text{Lateral Weight Transfer} = \frac{W \times g\text{'s} \times H}{\text{Gravity} \times T}$$

You can reduce the equation by factoring in a 1.0 *g* cornering force, the amount equal to the force of gravity, and then the formula would be:

$$\text{Lateral Weight Transfer} = \frac{WH}{T}$$

For our example, let's say our car has a center of gravity height of 20 inches. With a track of 60 inches,

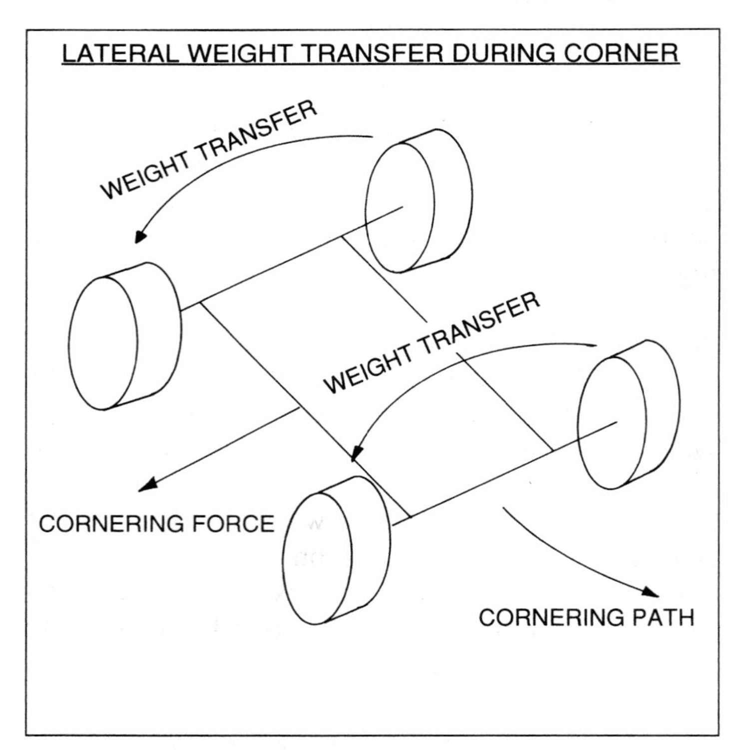

Figure 2-2. When a car goes around a corner, some of its weight transfers from the inside tires to the outside tires. This causes the weight on the outside tires to increase while the weight on the inside tires decreases.

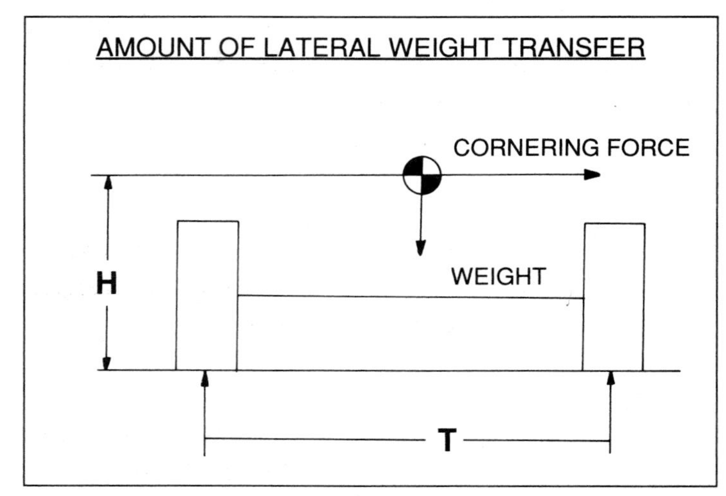

Figure 2-3. The amount of lateral weight transfer is dependent on the weight of the car, the magnitude of the cornering force, the height of the center of gravity (H), and the track width (T).

Chart 2-3

Tire Location	Static Weight On Tire	Lateral Weight Transfer	Weight On Tire During Cornering	Traction Available
Left Front	1050	-500	550	730
Right Front	450	+500	950	970
Left Rear	1050	-500	550	730
Right Rear	450	+500	950	970
Totals	3000		3000	3400

the weight transfer for a 3000-lb. car at 1.00 *g* cornering force would be:

$$\text{Lateral Weight Transfer} = \frac{3000 \times 20}{60} = 1000 \text{ lbs.}$$

This means 1000 lbs. of load will be transferred from the inside tires to the outside tires during cornering. On a car with equal front-to-rear weight distribution, this lateral weight transfer would be split evenly to 500 lbs. on each axle. If the car turns left, the left-side tires will lose 500 lbs. each and the right-side tire will gain 500 lbs. At these tire loadings, the traction available would be 3160 lbs. To see where we got that total, look at Chart 2-2.

Again, the traction available was calculated by using the tire performance chart, Figure 1-1 on page 2, as it will be for all of the charts in the following examples. Now, to find the cornering force, take the total traction available in Chart 2-2 and divide it by the total weight on the tire when cornering, or:

$$\text{Total Cornering Force} = \frac{3160}{3000} = 1.05\ g\text{'s}$$

As can be seen, cornering power was decreased (from 1.13 to 1.05 g's) because of the lateral weight transfer due to the cornering loads.

Example Three

One way to help equalize the weight on the tires during cornering is to *preload* the inside tires. This is done by moving some of the weight from the right side of the car to the left side of the car. This will obviously only work on a circle track, where the cars only turn left. In this case, assume the car has the following specifications:

Front End Weight:	**50%**
Left Side Weight Bias:	**600 lbs.**
Load Transfer From Cornering:	**1000 lbs.**

Under these conditions, the tire loadings and traction values would be as shown in Chart 2-3. Again, to find the total cornering force, plug the total traction available and total weight into the formula:

$$\text{Total Cornering Force} = \frac{3400}{3000} = 1.13\ g\text{'s}$$

This shows how increasing the weight on the left side (preloading) is effective in equalizing tire loadings during cornering, so that all the tires equally share the load and produce the maximum in cornering force.

Example Four

To see the effect of having a front-heavy car, let's see what happens when we make the front-end weight 60% with no left-side weight bias and our normal 1000-lb. weight transfer from cornering. With the same 3000-lb. car, there would be 1800 lbs. on the front tires and 1200 lbs. on the rear tires. The static weights, cornering weights and traction available would be as shown in Chart 2-4. Plugged into the formula, it would look like:

$$\text{Total Cornering Force} = \frac{3130}{3000} = 1.04\ g\text{'s}$$

Note that this is the average cornering force.

Chart 2-4				
Tire Location	Static Weight On Tire	Lateral Weight Transfer	Weight On Tire During Cornering	Traction Available
Left Front	900	-600	300	500
Right Front	900	+600	1500	1250
Left Rear	600	-400	200	380
Right Rear	600	+400	1000	1000
Totals	3000 lbs.		3000 lbs.	3130 lbs.

This total is misleading, because if you look at just the front-end weights and traction forces in Chart 2-4, you see that there is 1750 lbs. of traction for pulling the front-end weight of 1800 lbs. around corners. The front-end cornering force would then be:

$$\text{Front Cornering Force } \frac{1750}{1800} = .97\ g\text{'s}$$

At the rear, as shown in Chart 2-4, there is 1380 lbs. of traction to pull 1200 lbs. of weight around corners. This means the rear cornering force will be:

$$\text{Rear Cornering Force} = \frac{1380}{1200} = 1.15\ g\text{'s}$$

One method to equalize the weight on tires during cornering on oval tracks, where the car only turns left, is to preload the inside tires. Weight is moved from the right side of the car to the left side of the car, so that when weight is transferred, the loading on the outside tires will be less. See Example Three.

This analysis shows that the car in this example will not only corner slower than one with equal front-to-rear weight distribution, but it will also understeer in the corners. If the front traction is not able to pull the front weight as well as the rear traction can pull the rear weight, the front-end won't stick as well as the rear-end. This causes the car to understeer in the corners and to wear the right front tire faster than normal.

Even more important is to note that although the total traction will permit the car to corner at 1.04 *g*'s, the front-end won't corner at over .97 *g*'s. This means that because of understeer, the car can only corner at .97 *g*'s. This is considerably less than the 1.13 *g*'s shown in Example Three.

Example Five

This example is a combination for Examples 2, 3 and 4. Our interest with this combination is to see if the use of left-side weight bias will improve the cornering power of a front-heavy car and if it will solve the understeer problem. The static and cornering weights as compared to traction forces are listed in Chart 2-5. Plugged into the formula, the values are:

$$\text{Total Cornering Force} = \frac{3370}{3000} = 1.12\ g\text{'s}$$

This shows that the total cornering force is almost as good as in Example Three, but not quite as good. Also, if we look at the front-to-rear distribution of cornering forces separately, we can see that the car will still understeer. To do that, take the front and

Chart 2-5				
Tire Location	**Static Weight On Tire**	**Lateral Weight Transfer**	**Weight On Tire During Cornering**	**Traction Available**
Left Front	1200	-600	600	750
Right Front	600	+600	1200	1100
Left Rear	900	-400	500	700
Right Rear	300	+400	700	820
Totals	3000 lbs.		3000 lbs.	3370 lbs.

rear totals from Chart 2-5 and work them into the equation separately:

$$\text{Front Cornering Force} = \frac{1850}{1800} = 1.03\ g\text{'s}$$

$$\text{Rear Cornering Force} = \frac{1520}{1200} = 1.27\ g\text{'s}$$

As was the case in Example Four, the car will only be able to corner at 1.03 *g*'s because this is all the front-end traction the car has available. This value is still considerably lower than the 1.13 *g*'s shown in Example Three.

Example Six

The parameters for this exercise are the same as Example Five, except the chassis is *wedged* by adding 200 lbs. of weight to the right rear tire. Using a chassis wedge is a common method used to cure understeer. Wedging is accomplished by preloading the left front or right rear spring. When 200 lbs. of wedge is added to the right rear, the weight on the left front will also increase about 200 lbs., with a reduction in the weight on the right front and left rear of a similar 200 lbs. Assume our car has the following specifications:

Wedge Weight: 200 lbs.

Front End Weight: 60%

Left Side Weight Bias: 600 lbs.

Load Transfer from Cornering: 1000 lbs.

Taking this data and putting it into Chart 2-6 reveals that there is 3220 lbs. of traction available. Once again, take the data from Chart 2-6 and work it

A car that is front-heavy will have less traction available from the front tires to pull the front-end weight around the corner, while increasing the amount available at the rear to pull the rear-end weight around the corner. If the front traction is not able to pull the front-end weight as well as the rear traction pulls the rear-end weight, the front-end of the car won't "stick" as well, and understeer as a result. See Example Four. Photo by Michael Lutfy.

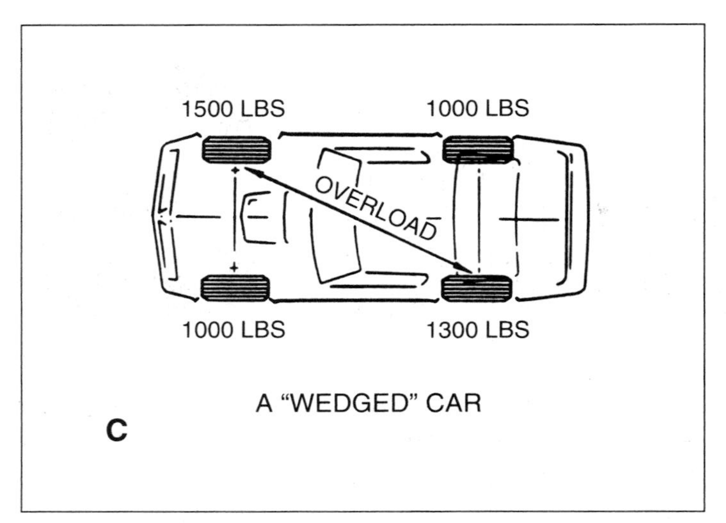

Figure 2-4. Using a chassis wedge is a common method used to cure understeer on circle track cars. Wedging is accomplished by preloading the left front or right rear spring. When 200 lbs. of wedge is added to the right rear, the weight on the left front will also increase about 200 lbs., with a reduction in the weight on the right front and left rear of a similar 200 lbs.

into the formula. Work the formula for the total cornering force, then just for the front and rear, as in Example Five.

$$\text{Total Cornering Force} = \frac{3350}{3000} = 1.12\ g\text{'s (average)}$$

$$\text{Front Cornering Force} = \frac{1900}{1800} = 1.06\ g\text{'s}$$

$$\text{Rear Cornering Force} = \frac{1450}{1200} = 1.21\ g\text{'s}$$

This example shows that the car will still understeer in the turns, but it will do it less, and the car will go around corners faster. The front traction is still weak, but wedging the chassis increased the car's front cornering force from the value shown in Example Five, 1.03 *g*'s to 1.06 *g*'s. This is a significant improvement, but it is still far less than the total 1.13 *g*'s we showed in Example Three.

SUMMARY

The numbers shown above should only be used to study the concepts. Each tire has its own performance curve, so you can't use these numbers to set up your car. However, analysis of these numbers does demonstrate some important guidelines:

1. The best cornering power is available when front-to-rear weight distribution is equal, assuming the tire size is equal both at the front and rear.
2. Left-side weight bias increases cornering power for oval track cars.
3. Cars that have front-end weight bias (heavier in front) will tend to understeer while cornering.
4. Wedging the chassis can reduce understeer in the turns and produce faster cornering.
5. In general, the best cornering power will result when all four tires are equally loaded during cornering.
6. Chassis tricks like left-side weight bars and wedging will only work on cars that turn left, like circle track cars. ■

Chart 2-6

Tire Location	Static Weight On Tire	Lateral Weight Transfer	Weight On Tire During Cornering	Traction Available
Left Front	1400	-600	800	900
Right Front	400	+600	1000	1000
Left Rear	700	-400	300	500
Right Rear	500	+400	900	950
Totals	3000		3000	3350

ROLL ANGLE & ROLL FORCE DISTRIBUTION

When a car goes around a corner it will roll towards the outside of the turn, which adversely affects handling. This is called *body roll*, and the amount that it rolls is called the *roll angle*. Various means are available to control the amount of roll angle and to minimize its negative effects on handling.

Resistance to body roll can be achieved at the front of the car, at the rear of the car, or at both the front and the rear. By deciding how much of the roll resistance is on the front and on the rear, you can control the understeer and oversteer characteristics of your car.

ROLL ANGLE

When a car rolls, the tires change their camber angle to the track surface (Figure 3-1). Since a tire develops its maximum traction when it runs perpendicular to the track, this positive camber angle results in less cornering power. Less roll angle results in less positive camber, so a car will corner faster if the roll angle is kept small.

Different suspension geometry factors such as roll center height, swing arm length, the height of the knuckle, the length of the control arms and the positions of the control arms all contribute to the amount of camber change that is realized for a given amount of roll angle. (All of these relationships will be discussed in future chapters.) As a practical matter, however, it is difficult to get a front-end geometry to work correctly when the negative camber gain is over about 3/4 degree per degree of body roll. This means that if the car rolls at a 4-degree angle, the outside tire will decamber 3 degrees, so the outside tire will lose 1 degree of camber in relation to the track.

Negative Camber

You can compensate for this loss of camber by setting the car up with *static negative camber*, which

If a car has a large roll angle, the tires will not be perpendicular to the ground, so they will not provide the maximum cornering power.

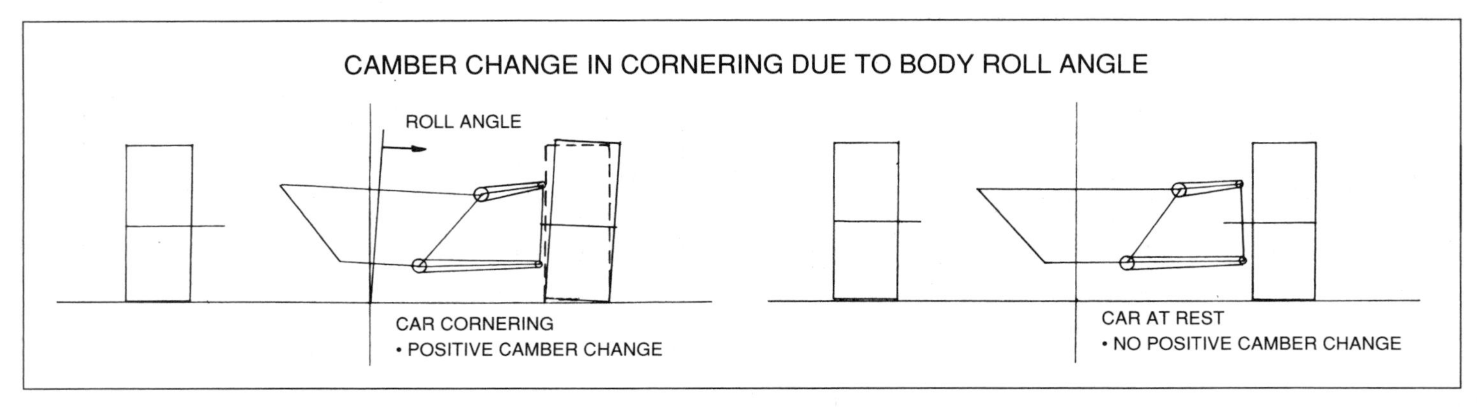

Figure 3-1. When a car rolls due to the cornering force, the tires usually roll with the car and develop a positive camber angle to the ground.

is the amount of negative camber with the car at rest. By doing this, you help to keep the outside front tire perpendicular to the track, even if there is excessive body roll.

Using excessive static negative camber can lead to problems, however. For most street applications, the maximum is about 1.0 degrees, or else the insides of the tires will wear. For competition, static negative camber settings of 2 or 3 degrees are often used. Tire temperatures can be used to optimize the amount of negative camber for most applications (see Chapter 16).

Solutions

Because there are limits to the amount of negative camber that can be used, it is important to control the amount of roll angle. The roll angle can be controlled to varying degrees with the following elements of chassis design and setup.

Center of Gravity Height—It is easy to see why a lower center of gravity will result in less roll angle. Most cars are already as low as is practical, so changing the height of the center of gravity is not always an available means of controlling the roll angle on a given car.

Roll Center Height—As can be seen in Figure 3-2, raising the suspension roll center will reduce the roll angle. Since the roll center height is an integral element of the total suspension geometry picture, we will discuss its effects in Chapter 7.

Track Width—Because the lateral spring base is proportional to the track width, a wider track dimension will reduce the roll angle. As was the case with center of gravity height however, most cars already have as wide a track as practical. This means that for any given car, we can not expect to cause much of a reduction in the roll angle by increasing the track dimension.

Cornering Force Amount—As can be seen from the GTO in the photo on page 13, more cornering force will result in more roll angle. If you want to go around corners as fast as possible, you will have ever-increasing cornering forces, and therefore ever-increasing roll angles. For example, if a street-driven

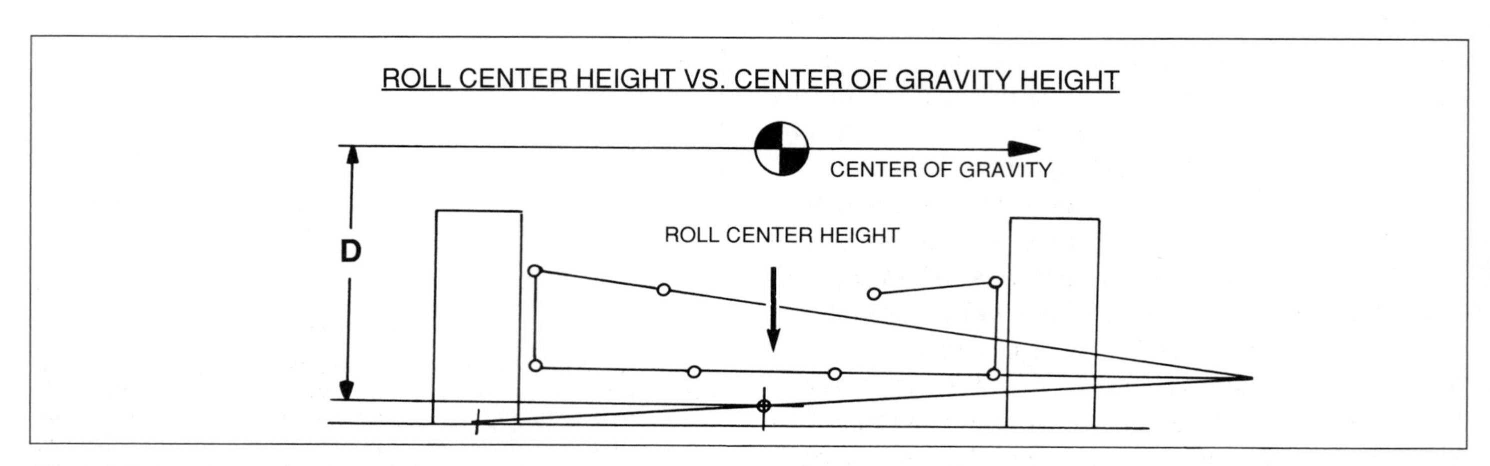

Figure 3-2. A car's roll angle is dependent on the distance between the height of the center of gravity and the height of the front and rear roll centers. The greater this distance, the greater the roll angle for any given cornering force.

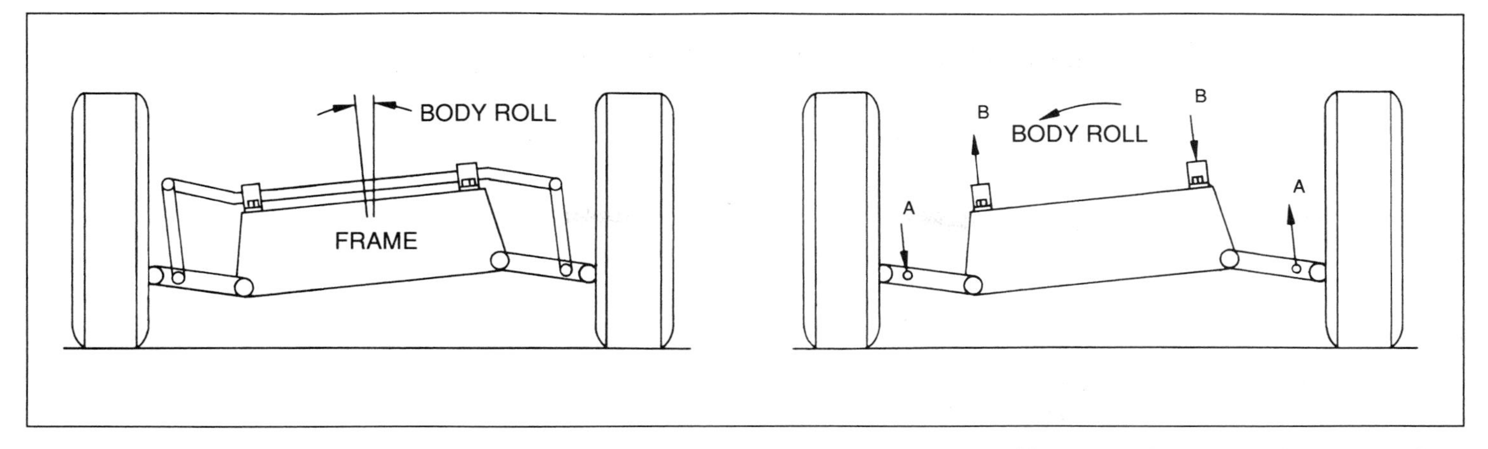

Figure 3-3. The left drawing shows how an anti-roll bar is twisted when the body rolls in a turn. This creates forces at the four points where the bar is attached to the vehicle. The forces are shown in the right drawing. Forces A on the suspension increase weight transfer to the outside tire. Forces B on the frame resist body roll. The effect is a reduction of body roll and an increase in weight transfer at the end of the chassis which has the anti-roll bar. Because the total weight transfer due to centrifugal force is not changed, the opposite end of the chassis has reduced weight transfer.

car with street tires could corner at .75 *g*'s, it might have a roll angle of 3 degrees. This same car on race tires might corner at 1.00 *g*'s. If it did, the roll angle would increase to 4 degrees. This means that cars that corner faster will need more roll stiffness to control the roll angle.

Roll Stiffness

The best way to control camber changes caused by body roll is to limit the roll angle by changing the *roll stiffness* of the suspension. The two most common means of controlling the roll stiffness on any given car are via the springs and the stabilizer bars.

Spring Rates—Increasing the spring rates will reduce roll angle. Unfortunately, raising the spring rates can also change other aspects of the car's handling. As an example, if a car had a front spring rate of 700 lbs.-inch and a roll angle of 2 degrees, and you wanted to reduce the roll angle to 1 degree, you'd need to install 1400 lbs-in. front springs. This would double the roll resistance. But increasing the spring rates this much would also upset the ride motions and cause the car to understeer. Springs are discussed in greater detail in Chapter 5.

Stabilizer Bars—The best way to increase roll stiffness is to increase the size or effectiveness of the *stabilizer bars*, which are sometimes called *anti-roll bars*. If a car is to roll, one wheel will be up in compression and one wheel will be in drooping down. Stabilizer bars limit the roll angle of a car by using their torsional stiffness to resist the movement of one wheel up and one wheel down. Connecting both wheels to each end of a stabilizer bar causes this motion to twist the bar (Figure 3-3). The stiffer the bar, the more resistance to body roll it can provide. Since the forces that cause the car to roll are being absorbed by the stabilizer bar, and these forces are fed into each lower control arm, the outside tire loadings will increase as the stabilizer bar twists. The stiffness of a stabilizer bar increases very quickly as its diameter is increased. The stiffness is a function of the diameter to the 4th power, or:

$$\text{Stiffness} = D^4$$

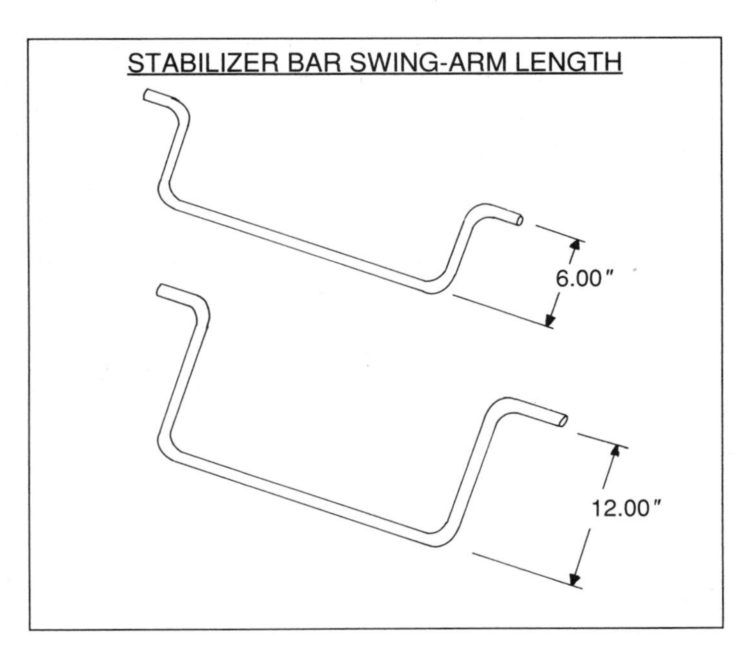

Figure 3-4. The effectiveness of a stabilizer bar is dependent on the length of the swing-arm as well as its diameter. The longer the swing-arm length, the less force the bar can provide with the same amount of movement at its end. For example, a stabilizer bar with a swing-arm length of 6 inches will produce twice the amount of roll stiffness as a bar with a 12-inch swing-arm length.

Chart 3-1

Tire Location	Static Weight	Lateral Weight Transfer	Weight w/ Cornering Load	Lateral Weight Transfer w/Front Stabilizer Bar	Weight w/ Front Bar	Traction Available
LF	1050	-500	550	-400	150	320
RF	450	+500	950	+400	1350	1200
LR	1050	-500	550	000	550	730
RR	450	+500	950	000	950	970
Totals	3000		3000		3000	3220

This means that a 1 1/4-inch diameter stabilizer bar is 2.44 times as stiff as a 1.00-inch diameter stabilizer bar. But, the stiffness of the bar must be properly transmitted into the chassis to do any good. The length of the arms that feed the stabilizer bar loads into the chassis have a dramatic effect on how much roll stiffness a given bar can produce on the chassis. The longer the bar, the less effective it will be. For example, 6.00-inch long stabilizer bar arms will produce twice as much roll stiffness as 12.00-inch long arms (Figure 3-4). Also, the total roll stiffness of a given stabilizer bar is dependent on the stiffness of the frame mounting, the stiffness of the arms, the stiffness of the drop links, and where the drop links connect to the lower control arms.

The effectiveness of a stabilizer bar is also dependent on how well the bar is mounted to the frame and to the control arms. Any lost motion at these connections will result in a loss of bar effectiveness. Make sure you use high-quality, rigid materials for the construction of links and brackets.

ROLL FORCE DISTRIBUTION

By varying the size and effectiveness of the front stabilizer bar vs. the size and effectiveness of the rear stabilizer bar, it is possible to change the understeer/oversteer characteristics of a car. Since the forces resisting the roll of the car are fed to the outside tires, it is possible to decide whether the front outside tire or the rear outside tire will absorb most of these forces. If a car has understeer, too much load is on the front outside tire. By increasing the effectiveness of the rear stabilizer bar, some of this load can be transferred during cornering to the outside rear tire. Doing this will eliminate the understeer, because the front and rear outside tires will be more equally loaded.

The following examples show how transferring some of the forces resisting body roll can be fed into the rear tires to eliminate understeer. These examples use the same tire curve as shown in Figure 1-1

Chart 3-2						
Tire Location	**Static Weight**	**Weight Transfer**	**Weight with Cornering Load**	**Weight Transfer w/Front & Rear Stabilizer Bars**	**Weight w/ Front & Rear Bars**	**Traction Available**
LF	1050	-500	550	-200	350	550
RF	450	+500	950	+200	1150	1100
LR	1050	-500	550	-200	350	550
RR	450	+500	950	+200	1150	1100
Totals	3000		3000		3000	3300

(Chapter 1 page 2) and use the same procedures as explained in Chapter 2. These examples are for a circle track car that only turns left. On cars that turn left and right, the sample principles apply, but in opposite directions.

Example One

For the figures in Chart 3-1 on p. 16, we assumed we had a car with the following specifications:

Car Weight: 3000 lbs.

Front End Weight: 50%

Cornering Load Transfer: 1000 lbs.

Left-Side Weight Bias: 600 lbs.

Roll Stiffness (Front Only): 400 lbs.

Now, using the information in Chart 3-1, plug it into the formula for total cornering force, which was covered in Chapter 2.

$$\text{Total Cornering Force} = \frac{\text{Traction}}{\text{Weight}}$$

$$\text{Total Cornering Force} = \frac{3220}{3000} = 1.07\ g\text{'s (average)}$$

To find the front cornering force only, take the totals of both front tires only from Chart 3-1 and plug into the formula:

$$\text{Front Cornering Force} = \frac{1520}{1500} = 1.01\ g\text{'s}$$

And for the rear cornering force:

$$\text{Rear Cornering Force} = \frac{1700}{1500} = 1.13\ g\text{'s}$$

These figures indicate understeer, because there is more cornering power available at the rear than at the front. This is what happens when you apply roll stiffness at the front only. Let's see what happens when you equalize the roll stiffness.

Example Two

For this example, the car specifications are the same as Example 1, except that instead of having 400-lbs. roll stiffness on the front only, we've split the roll stiffness to 200 lbs. on the front and 200 lbs. on the rear by using both a front and rear stabilizer bar.

Again, the load transfer and traction available in Chart 3-2, is taken and plugged into the formula for cornering force to get the force in *g*'s, or:

$$\text{Total Cornering Force} = \frac{3300}{3000} = 1.10\ g\text{'s.}$$

And now, to find out how the cornering force balances out with equal front and rear stabilizer bars,

Rear roll stiffness changes using a rear stabilizer bar are the primary means of balancing the understeer/oversteer characteristics of road-race and high performance street cars. Photo by Michael Lutfy.

take the information from Chart 3-2 for the fronts and rears separately and plug them into the formula:

$$\text{Front Cornering Force} = \frac{1650}{1500} = 1.10\ g\text{'s}$$

$$\text{Rear Cornering Force} = \frac{1650}{1500} = 1.10\ g\text{'s}$$

In this example, there is no understeer because the cornering forces at both the front and rear are equal. When comparing the two examples, note that the total available *g*-forces are increased when a rear stabilizer bar is used to control half of the roll angle. More importantly, the front *g*-forces are increased by 9% and the car no longer suffers from understeer. Balancing a car's stabilizer bars is as an important aspect of chassis tuning as balancing the springs and the static weights.

SUMMARY

All of the preceding information can be summarized into the following:

1. Increasing the rear roll stiffness reduces understeer.

2. As a larger rear stabilizer bar will reduce understeer because of its ability to increase the dynamic weight on the outside rear tire, it will also work on cars that turn in both directions.

3. Because left-side weight bias and weight wedging will not work on a road-race car or a street-driven car, rear roll stiffness changes from a rear stabilizer bar are the primary means (although not the only method) of balancing the understeer/oversteer characteristics of these cars. ■

BUSHINGS & DEFLECTIONS

Even the best suspension geometry and suspension alignment is only effective if the suspension components do not change shape or position. Suspension pieces in general look strong and rigid, but these pieces must handle loads in the 1000-5000 lbs. range, so they do bend and deflect. Under high performance driving conditions, the wheels bend, the knuckles bend, the control arms bend and the frame bends. The worst source of deflection on a production car is the suspension bushings. Suspension bushings are deceptively simple devices. Insignificant as they may seem, they have a very important effect on your car's handling. Therefore, the construction and quality of your bushings deserves close attention.

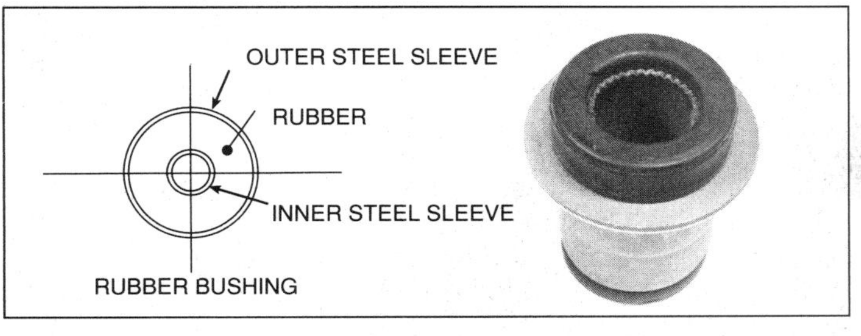

Figure 4-1. Rubber suspension bushings are used on most production cars today. They consist of an inner sleeve, outer sleeve, and some form of material (almost always rubber in the case of production cars) separating the two. Rubber bushings are used primarily for cost, better road noise isolation and they don't require lubrication.

RUBBER BUSHING DEFLECTION

Most bushings consist of an inner sleeve, an outer sleeve and some form of material separating the two. In most modern production bushings, the material bonded between the sleeves is rubber. During the 1950's, factory chassis designers were attracted to rubber suspension bushings because they offered three advantages over earlier steel-on-steel versions. The rubber bushings were (1) cheaper to build, (2) offered better isolation from the jolts of the road and (3) didn't require lubrication. Also, because the *durometer* (hardness) of the rubber could be tailored for specific chassis characteristics, the engineers gained a new design flexibility they didn't have with steel bushings.

If rubber bushings have all of these advantages, why are so many car enthusiasts replacing their bushings with ones made of other materials? Obviously, there must be driving situations in which the rubber bushings do not perform as well as they should.

Even high performance street cars, like this Pontiac Trans Am, are subject to the ills of rubber bushing deflection. Aftermarket replacement bushings are available to replace the stock units, the reasons for which are explained throughout this chapter.

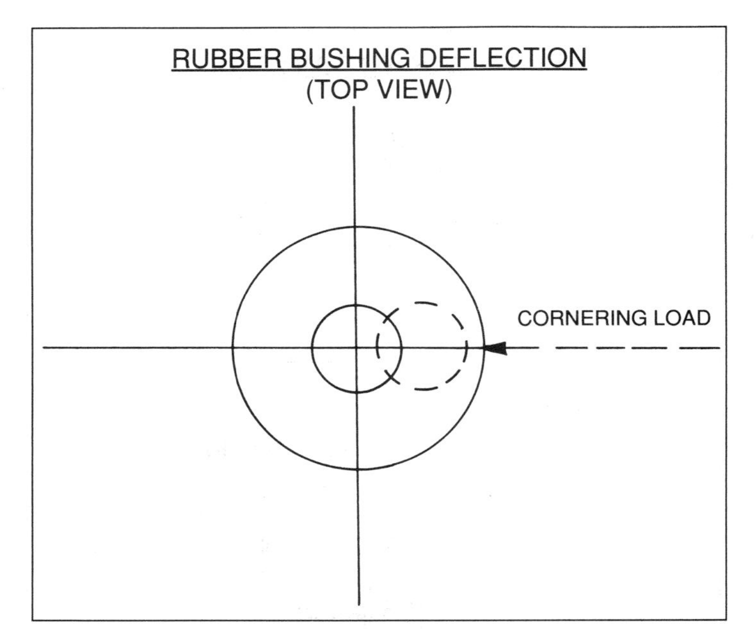

Figure 4-2. For most street driving, rubber bushings work very well. But under high performance driving conditions, the rubber deflects, allowing the inner sleeve to move toward the outer sleeve, which changes the location of the control arm.

Under most driving conditions, rubber bushings are the best choice. However, high performance driving demands less deflection, which a rubber bushing is not capable of providing. The excessive deflection of rubber bushings can have several adverse effects on high performance handling (Figure 4-2).

Loss of Camber Control

A tire generates its maximum cornering power when it is perpendicular to the road surface. If the suspension bushings deflect when they are loaded by high cornering forces, the tire is forced to positive camber—and cornering forces are reduced (Figure 4-3). This explains why setting static negative camber helps cornering power. By aligning the suspension with negative camber when the car is at rest, you anticipate the positive camber caused by rubber bushing deflection during cornering. If the suspension bushing material resists deflection, there is less loss of camber control.

The loss of cornering power due to rubber bushing deflection is a problem on cars with independent front suspension and on cars with independent rear suspension. The need for a better bushing material is especially important to Corvette owners because camber loss at the rear will cause oversteer (Figure 4-4).

Front Deflection Steer

Rubber suspension bushing deflection can also have a dramatic effect on the steering characteristics of your car. The steering linkage consists of rigid links and joints, so there is little deflection when these parts are loaded by high cornering forces. The control arms, however, are mounted in

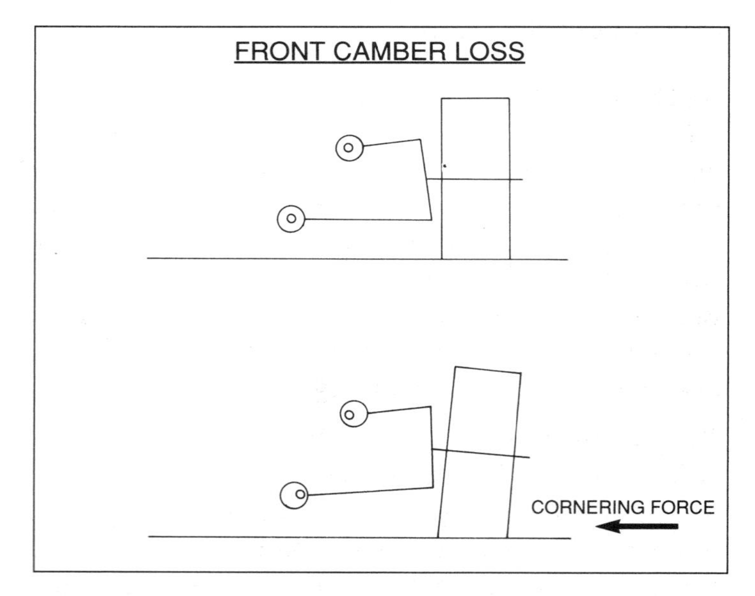

Figure 4-3. When the front control arm bushings deflect, the control arms can move in relation to the frame. The cornering loads cause this deflection to result in positive camber, which reduces and distorts the tire patch. The net effect is a loss of cornering power at the front.

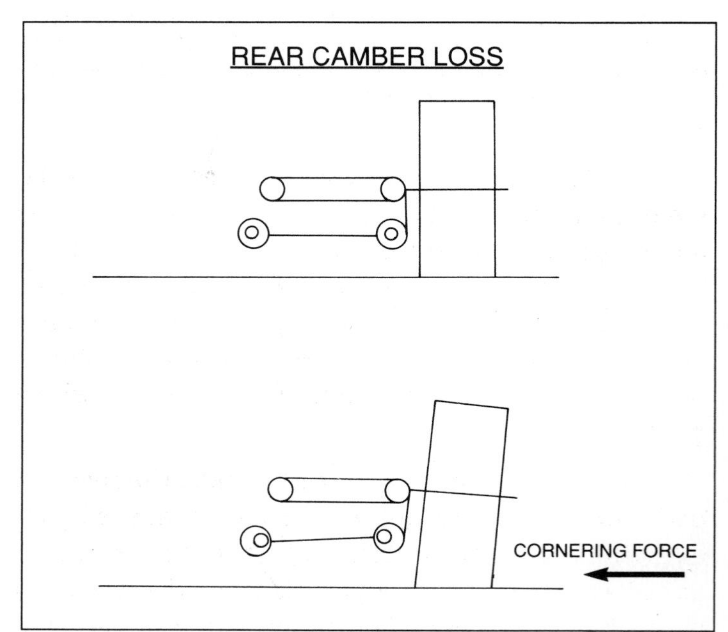

Figure 4-4. If a car has independent rear suspension with rubber bushings, the deflection will again result in positive camber and a loss of traction at the rear wheels.

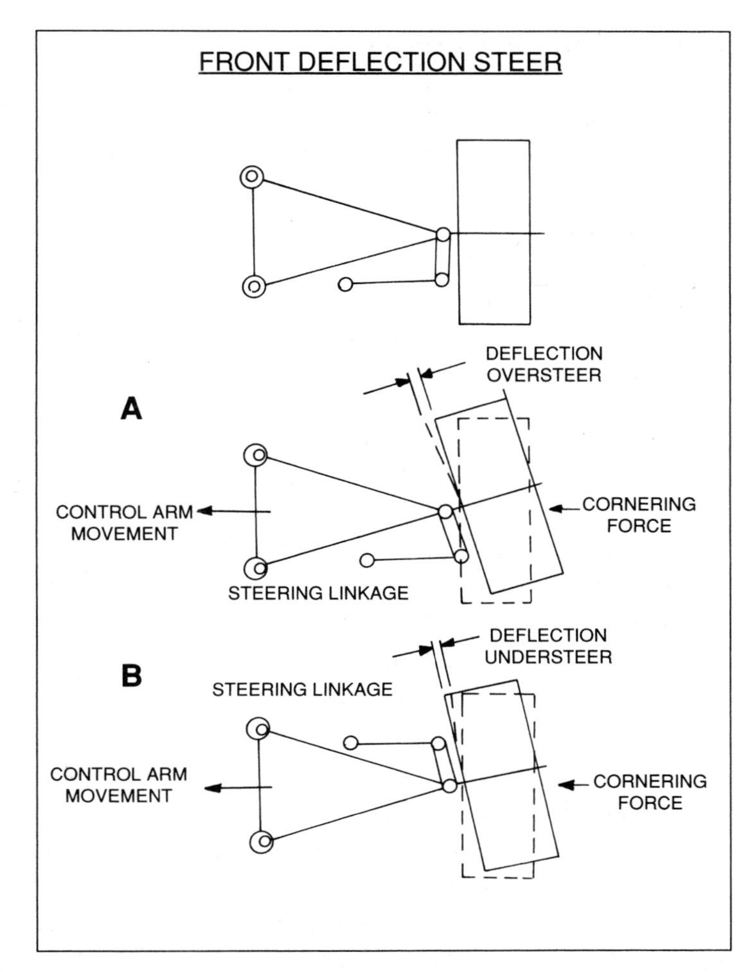

Figure 4-5. When the control arms move because of bushing deflection, they also cause a change in steering angle, because the steering rods do not move as the control arms move. If the steering linkage is behind the front axle (A), this deflection can cause oversteer, which is unstable during high performance driving. When the steering linkage is in front of the front axle (B), the result is deflection understeer, which is considered to be safer and easier to drive. This is why most late-model production cars have the steering linkage located in front of the axle.

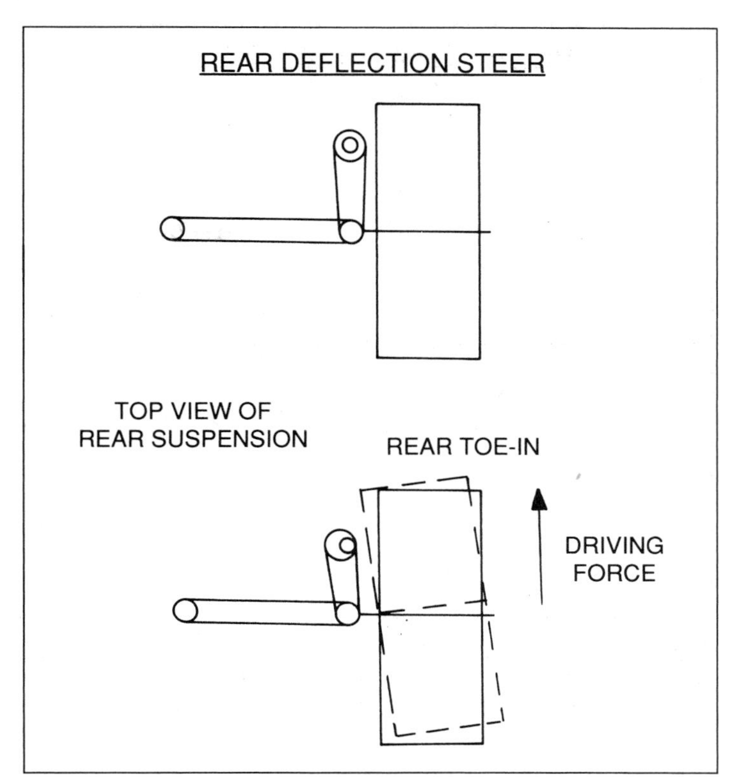

Figure 4-6. On a rear-wheel-drive car with independent rear suspension, rubber bushing deflection can cause torque steer, because the rear hubs can move during acceleration and/or braking. During acceleration, the rear wheel will toe-in (shown here) and cause understeer. The opposite, rear toe-out, occurs during braking. This change in the rear wheel alignment can result in handling that is unstable and unpredictable.

rubber. Thus, the control arms can deflect while the steering linkage stays in place. If the control arms move in and out, the steering knuckle will also move. The result is a change in *steering angle*. The driver will notice that the car feels "twitchy," because the actual steering angle changes as the bushings deflect, even though he is holding the steering wheel steady.

If the linkage is in front of the knuckle, the result is *deflection understeer*. If the steering linkage is located behind the front knuckle, the effect is *deflection oversteer*. Deflection oversteer is when the tire turns more than the driver asks. Deflection understeer is when the tire turns less than the driver asks. Deflection understeer is generally considered to be safer and more pleasant to drive, so most late-model automobiles have their steering linkages in front of the knuckles (Figure 4-5).

Rear Deflection & Torque Steer

Deflection steer can have a more dramatic effect on the handling of independent rear suspensions. If the suspension bushings permit changes in the rear tires' camber and toe, the car's directional stability will change over bumps and during hard cornering. Because the car's rear suspension must also absorb driving loads (which also deflect the bushings), changes in directional stability are common.

Torque Steer—If your car understeers under power and oversteers during braking, torque steer may be the culprit. Torque steer is when the steering angle of the driving wheels changes as the power and/or braking torques are applied. Under power, the rubber bushing at the front of the rear control arm compresses and the arm moves forward. This increases rear toe-in. More rear toe-in creates understeer,

Urethane bushings prevent deflections, but because urethane is a sticky plastic, the use of urethane bushings usually results in a suspension which is not free to move. This binding is made worse when the lubrication is forced out of the joint over time. Graphite-filled urethane does not solve the problem.

because the driving force aims the car toward the outside of the turn.

The opposite happens during braking. The bushing deflects and produces toe-out and oversteer. This varying condition—understeer under acceleration and oversteer during braking—doesn't promote driver confidence. A good-handling car should be predictable and controllable.

The first step in eliminating erratic handling is to reduce deflection in the control arm bushings (Figure 4-6).

STEEL-ON-STEEL BUSHINGS

There are several types of suspension bushings that will eliminate deflection. One obvious alternative is the old steel-on-steel bushings used by the factories before the development of rubber bushings. Steel-on-steel bushings require periodic greasing and close tolerances to operate properly.

Like all non-rubber bushings, they offer no isolation between the control arms and the chassis, so more road noise is transmitted to the passenger compartment. Steel-on-steel bushings are used on some race cars because road noise is not a concern.

URETHANE BUSHINGS

In recent years, several companies have offered urethane suspension bushings. These plastic bushings look great on the parts shelf. Unfortunately, the urethane used in the products we have evaluated is not a material well suited for suspension bushings.

Disadvantages

Depending on the application, suspension bushings must allow movement in different planes. The most common plane of movement is simple rotation. Rubber bushings allow rotation by the internal shear of the rubber itself. This means there is no sliding motion between any of the members. The rubber flexes to allow the inner and outer sleeves to rotate relative to each another. Since there is no sliding motion, there is no friction-caused wear and no need for lubrication. Because the rubber is molded to the inner and outer sleeves, there are no critical tolerances to maintain during manufacturing. This is one of the features that allows rubber bushings to be made inexpensively.

When a steel or urethane suspension bushing is used, the bushing material cannot deflect. There must be some sliding motion to permit rotation between the inner and the outer sleeve (Figure 4-7).

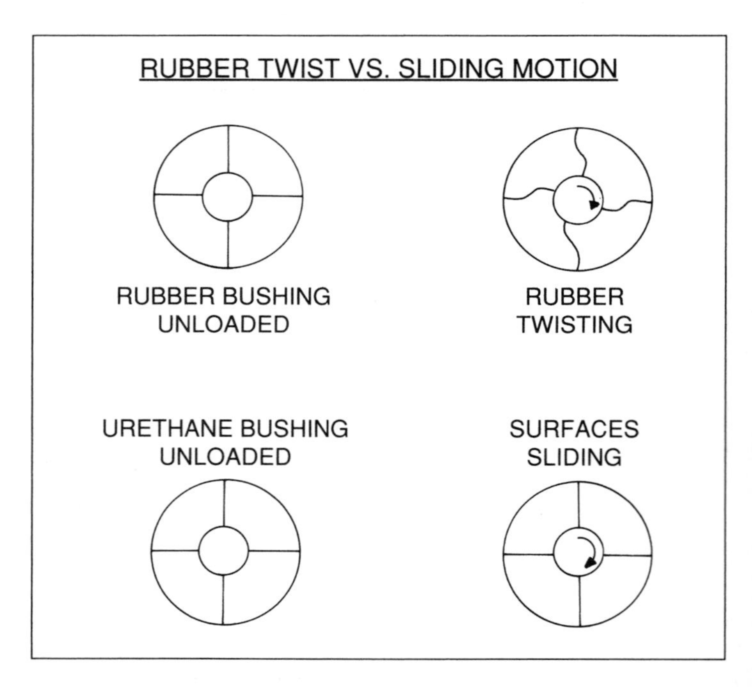

Figure 4-7. A rubber bushing allows the inner sleeve to rotate in relation to the outer sleeve by deformation of the rubber itself. Any hard bushing, like urethane, requires a sliding motion between the inner sleeve and the bushing material. This type of bushing is really a bearing, so slippery-type plastics are required.

Industrial-grade nylon bushings prevent bushing deflection. Their use requires machined steel support sleeves, so they cost slightly more than urethane bushings. Because they have grease fittings, they can last over 100,000 miles with bi-annual greasings.

Many people who buy aftermarket urethane bushings believe that the urethane behaves like hard rubber, but most urethane suspension bushings are so hard they have to be considered solid, because they offer little ability to absorb rotational shear within themselves. The only way a rigid suspension bushing can allow rotational movement is if one of the mating surfaces slides on the other. This sliding action demands close manufacturing tolerances and lubrication for satisfactory operation. The aftermarket urethane bushings we've tried did not meet these requirements.

Greasing the bushings before assembly works for a few weeks until the grease is forced out or washed away. Once the grease is gone, the urethane is again able to bind to the steel sleeves, and the driver hears squeaks and moans. Without lubrication, the urethane can stick to the steel, and the suspension does not move smoothly.

Urethane bushings could be customized to provide better service by hand-fitting the bushings to obtain proper clearances and installing grease fittings for regular lubrication. But, there is a better alternative.

NYLON BUSHINGS

Nylon bushings, when properly supported and lubricated, have proven themselves in many industrial applications. Nylon inserts are inexpensive and are available from any bearing supply house. These "nyliner" inserts require sleeves machined to close

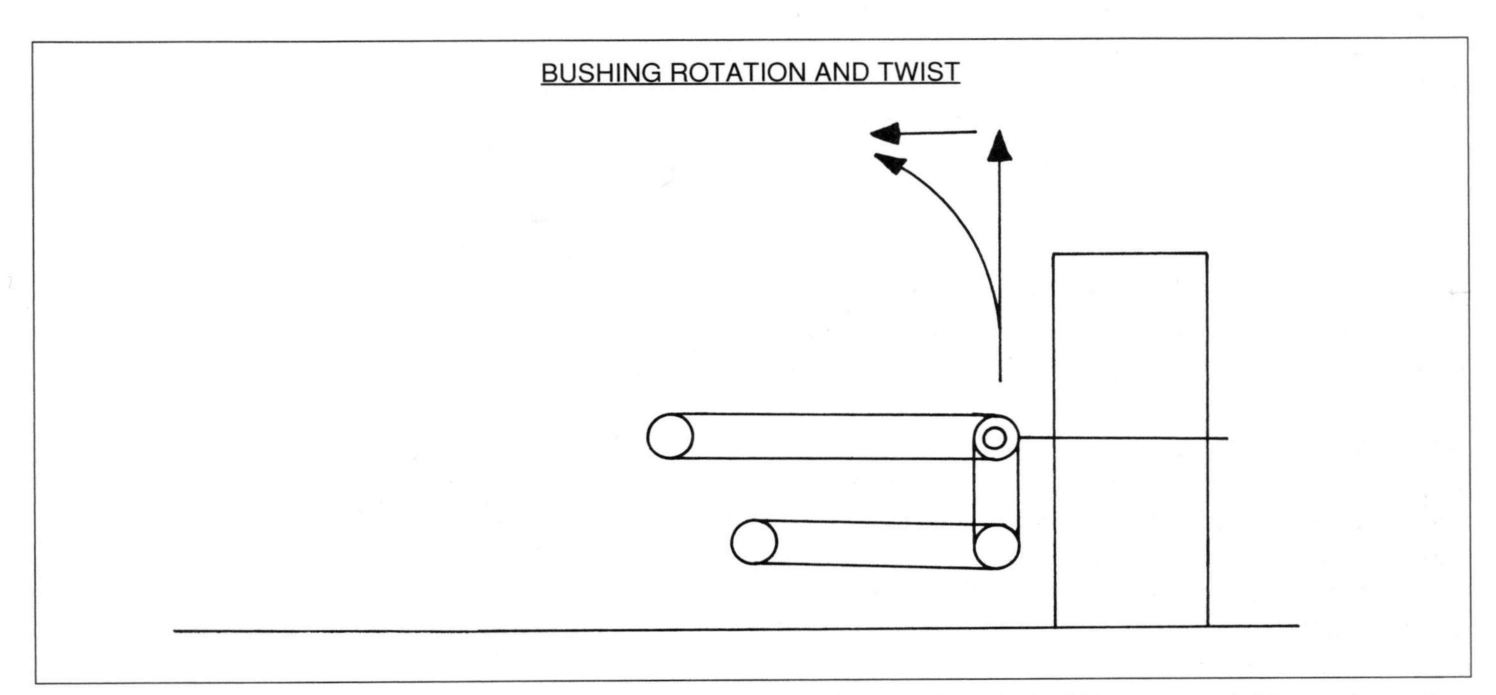

Figure 4-8. When a suspension member must move in rotation and twist at the same time, the bushing must permit this complex motion. Many rear suspension bushings are of this type, because they must allow the chassis and body to roll while at the same time allow the axle to move up and down. This diagram shows an early Corvette rear suspension, which has the same requirements.

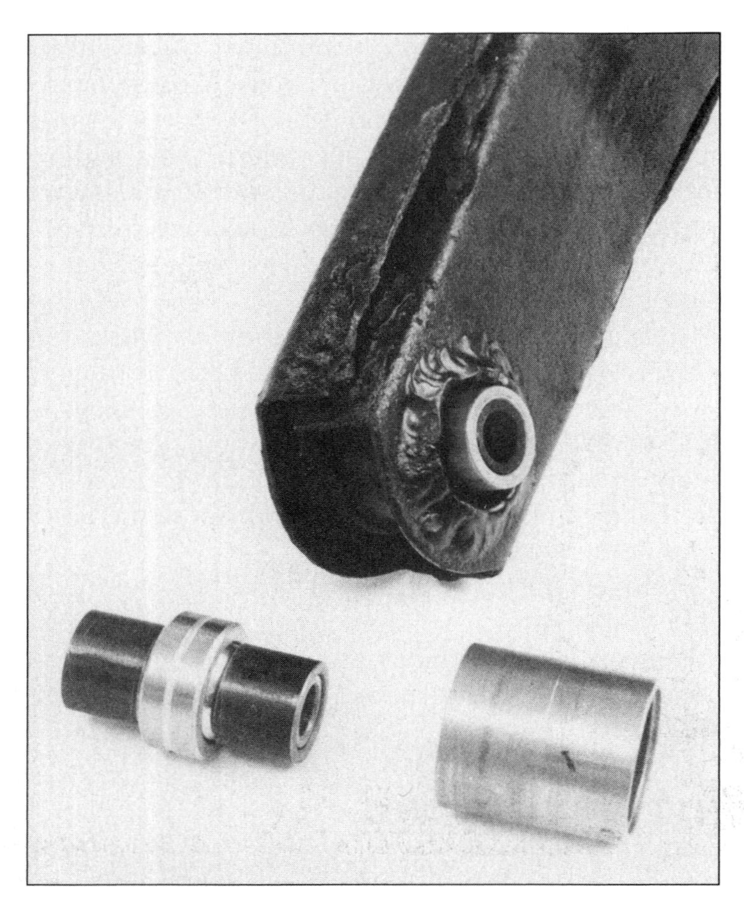

When the control arm motion requires rotation as well as twist, like on most rear suspension links, spherical bushings are needed to handle the complex motions. Using plain-type bushings in this application will result in a suspension that binds up.

tolerances for proper operation, so the total cost of installing nylon suspension bushings is slightly higher than urethane bushings. Compared to the cost of labor to remove and replace the control arms, however, the amount of increase is insignificant.

Properly made sleeves to support nyliner bearings should have zerk fittings for easy lubrication. I have run cars more than 100,000 miles with the same set of nyliner suspension bushings. I recommend the use of nylon and steel bushings for all applications where pure rotational motion is required. All bushings should have zerk fittings so they can be greased every six months.

SPHERICAL BEARINGS

Some suspension bushings are subject to a combination of rotational movement and angular twist. The bushings used between the frame and rear control arms on '63-'82 Corvettes are an example. Rubber bushings work in this case, because the rubber can deflect axially and rotationally at the same time. Simple urethane and nyliner bushings can not provide this combination of motions, and are not satisfactory for this use (Figure 4-8).

Spherical bearings can provide this combination of motions. Spherical bearings are expensive, however, so I recommend them only for this specific type of application. Many grades and levels of quality are currently available to the spherical bushing buyer. Testing has convinced me that only aircraft quality, hardened steel-on-hardened steel bearings will provide long-term durability. Considering the labor required to install the bushings, using lower quality, less durable bearings makes poor economic sense.

Spherical bearings are made to very close tolerances, so their supporting parts must also be made accurately. As a result, they are more expensive than cheap, molded parts.

RIDE & NOISE CHARACTERISTICS

If the suspension bushings you install are free to rotate without binding, the ride characteristics of your car will not change. But, if you eliminate the rubber from the suspension bushings, more road noise will be transmitted to the passenger compartment. I have found that nylon and spherical bearing suspension bushings will usually increase the impact harshness to the same degree as adding 5 psi to your car's tire pressure.

If you don't drive your car hard, or in competition, you probably do not need solid suspension bushings. However, if you can stand a little more impact harshness, installing suspension bushings will make your car handle like a race car. Both cornering power and steering response will improve greatly. ■

SPRINGS & SHOCKS 5

Springs and shocks are an integral part of any suspension system. But the total suspension system must be considered as a coordinated package, so just changing springs and/or shocks will not always give the desired results. What works on one car might not work on another car because of differences in their system design. The following aspects should be considered in choosing springs and shocks for your car.

SPRING LOAD & SPRING RATE

Spring load and spring rate are often confused. For the best understanding of how your suspension works, it is important to recognize the difference. *Spring load* is the amount of weight it takes to compress the spring to a given height, expressed in lbs. *Spring rate* is the amount of weight it takes to compress the spring one inch expressed in lbs-in. An example of this relationship for a spring is seen in Chart 5-1.

Note that the spring rate does not change as the spring is compressed, but the spring load does. In Chart 5-1, the spring compresses 1 inch for every 150 lbs. of load. The spring rate determines how much the spring will compress as the loading increases. The spring load determines how much weight the spring can support at a given height. For example, when the spring is compressed at 7 inches, the spring can support 1050 lbs. If one corner of your car is sagging, you don't need more spring rate, you need more spring load.

Spring Rate at the Wheel

Assume a car weighs 3000 lbs. and that it has equal weight distribution. This would mean it would have 750 lbs. of weight at each wheel. If we assumed an *unsprung weight* of 150 lbs. at each wheel, the *sprung weight* at each wheel would be 600 lbs. Unsprung weight includes parts of the car that are not supported by the springs, which would be the weight of the tire, wheel, knuckle, hub and one-half of the spring, shock and control arms. Sprung weight includes the weight of the body, frame, engine, transmission and one-half of the spring, shock and control arms. With a sprung weight of 600

Shocks and springs are an integral part of any suspension system. Matching the proper spring rate and load to suit driving and track conditions is a critical factor to racing success. However, the process is a bit more involved than just changing to different shocks and/or springs. Photo by Michael Lutfy.

Chart 5-1

Typical Rear Spring

Amount of Spring Compression	Spring Rate (lbs/in.)	Spring Load (lbs.)
1.00 inch	150	150
2.00	150	300
3.00	150	450
4.00	150	600
5.00	150	750
6.00	150	900
7.00	150	1050
8.00	150	1200

lbs. at each wheel, you'd need a spring load of 600 lbs. to hold the car at normal ride height. If the wheel travel was 4.00 inches up and 4.00 inches down, the spring rate at the wheel would need to be 150 lbs. per inch to absorb a 1.00 *g* bump. A one *g* bump is equal to the normal irregularities on a smooth road. This spring rate would provide full suspension travel, and would be considered soft by most standards. However, with high performance applications, there are many factors that will require the use of higher rate springs.

Wheel Rate vs. Spring Rate

This relationship often becomes confusing when enthusiasts try to actually measure the results. *Wheel rate* is the actual spring rate at the wheel as opposed to spring rate at the spring. It is not a simple relationship, so adding 100 lbs. of spring rate will not add 100 lbs. of wheel rate unless the spring is mounted directly on the axle. Anytime there are linkages such as control arms involved, you have to consider the *linkage ratios*. Since most modern cars use upper and lower control arms on their front suspension, we will use this example to demonstrate how linkage ratios work. Figure 5-1 shows a schematic of a typical front suspension having an upper and a lower control arm with the spring acting on the lower control arm. The formula for determining the wheel rate is:

$$\text{Wheel Rate} = \text{Spring Rate}\left(\frac{a}{b}\right)^2 \times \left(\frac{c}{d}\right)^2$$

A: Distance from control arm inner control arm pivots to the center point where the spring acts on the lower arm.

B: The length of the lower control arm from the ball joint to the inner pivots.

C: The distance from the lower ball joint to the front suspension instant center.

D: The distance from the center of the tire contact patch to the front suspension instant center.

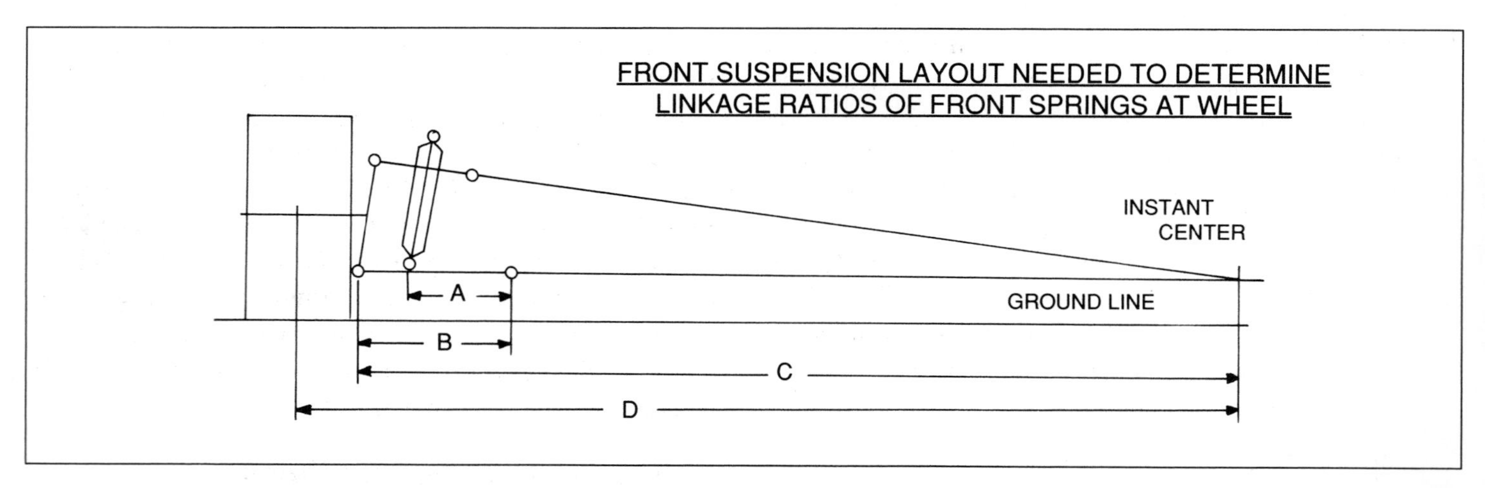

Figure 5-1. The spring rate at the wheel is not the same as the rate of the spring if there are any linkages involved. This shows a typical front suspension arrangement listing where the dimensions for the control arm and instant center point are located. To decipher what the coordinates are, see the text nearby.

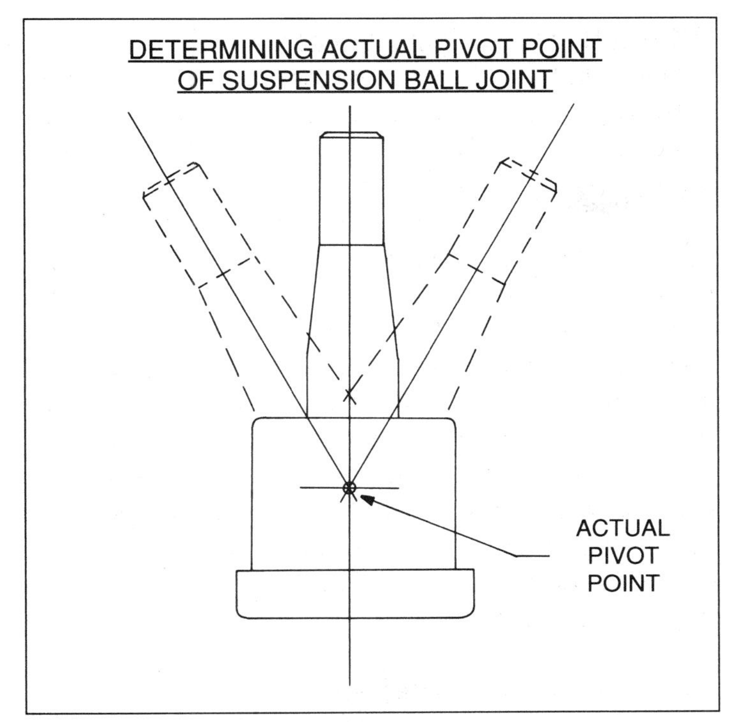

Figure 5-2. When making an accurate layout of a suspension system, it is important to know the exact pivot point of the ball joints. By moving the joint from one extreme to the other, it is possible to project this point. An even more accurate method would be to machine one side off the joint to measure what is inside.

Finding Instant Center—To accurately find the instant center of your front suspension, you need to make a scaled drawing based on measurements you have taken off your car with the car setting at ride height. This sounds like a lot of work, but you can do it in about an hour with a tape measure. It is not necessary to measure more accurately than 1/16 inch to get good results. Care should be taken, however, in determining the actual pivot point of the ball joints. This is difficult to measure since it is inside the joint. If you have a ball joint off the car, you can move it from one extreme to the other and eyeball what appears to be the actual pivot point. It is usually close to the flange or the end of the housing, so you can use this as a reference point in making the measurements on your car (Figure 5-2).

When you have made your front-end layout, you will have the A-B-C-D dimensions you need to fill in the wheel rate equation. For the purposes of explanation let's assume your dimensions are:

A = 9.0 in.

B = 14.0 in.

C = 90.0 in.

D = 94.0 in.

If the dimensions are plugged into the formula it would look like this for a 500-lbs-in. spring rate:

$$\text{Wheel Rate} = 500 \left(\frac{9}{14}\right)^2 \times \left(\frac{90}{94}\right)^2$$

Wheel Rate = 500 x $(.64)^2$ x $(.96)^2$
Wheel Rate = 500 x .41 x .92
Wheel Rate = 188 lbs-in.

Note that with this configuration the wheel rate is about 37% of the spring rate. This means that adding 100 lbs. of spring rate will only add 37 lbs-in. of wheel rate.

Using the same example, let's see what happens when the spring location on the lower control arm is moved outboard 1.00 inch. With this configuration:

A = 10 in.

B = 14 in.

C = 90 in.

D = 94 in.

$$\text{Wheel Rate} = 500 \left(\frac{10}{14}\right)^2 \times \left(\frac{90}{94}\right)^2$$

Wheel Rate = 500 x $(.71)^2$ x $(.96)^2$
Wheel Rate = 500 x .50 x .92
Wheel Rate = 230 lbs-in.

This wheel rate represents about 46% of the spring rate, so moving the spring acting point outboard 1.00 inch increased the wheel rate more than if another 100 lbs-in. of spring rate had been added.

Because of the effects of the linkage ratios, it is usually impossible to compare the spring rates of one car vs. the spring rates of another car unless all the linkage dimensions are known to be exact. From the above examples, it is also obvious that just using wheels with more or less offset will have a slight effect on the actual wheel rate.

Calculating Coil Spring Rates

If you have access to a spring rate checker, you can measure all of your springs. If you don't have a

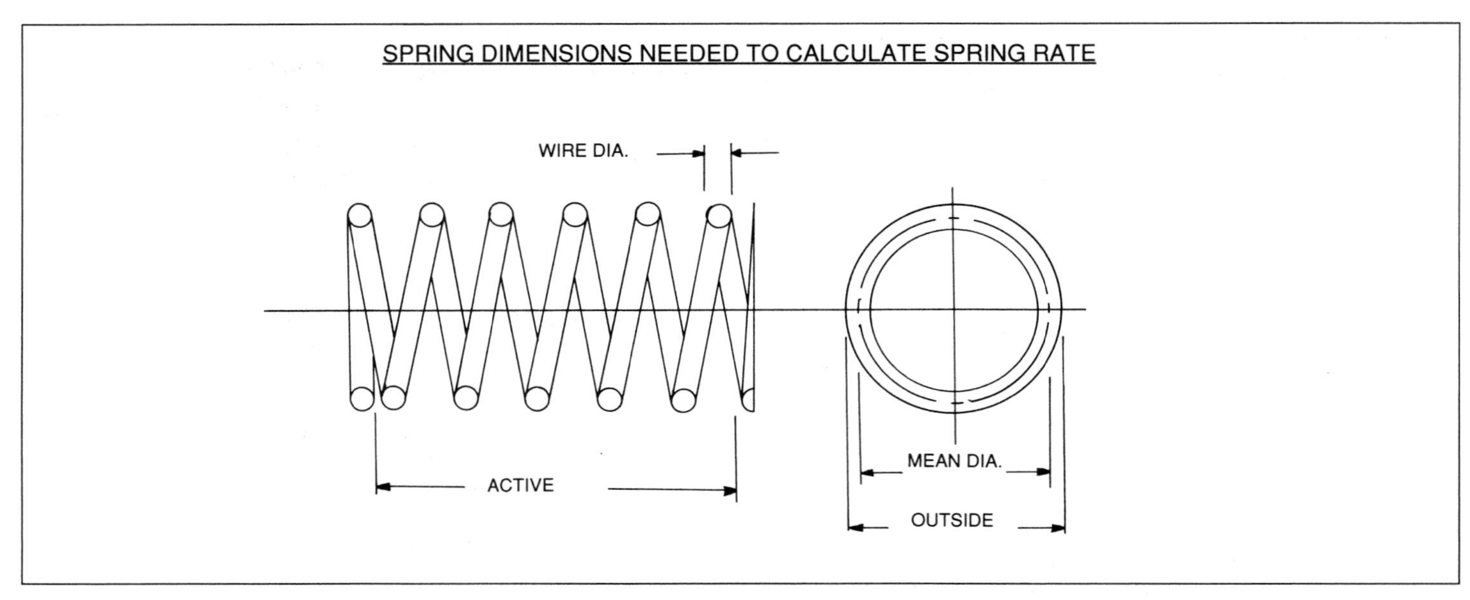

Figure 5-3. When calculating coil spring rates, it is necessary to determine the specific spring dimensions. This diagram shows where these dimensions are measured from for a typical coil spring.

checker, you can also determine the rate of your coil springs by measuring them and using this formula:

$$\text{Spring Rate} = \frac{Gd^4}{8ND^3}$$

G = Torsional modules for steel = 11.25 x 10^6 or 11,250,000

d = Wire diameter in inches

N = Number of active coils

D = Mean coil diameter in inches

8 = A constant for all coil springs

The G factor in the equation is always the same for coil springs made from steel. A titanium spring would require another factor (11.25 x 10^6 can also be written as 11,250,000).

The coil wire diameter (d) can be measured with a caliper. It will be the same for the whole length of the spring unless it is a spring of variable rate, in which case you can't put on number on it anyway. For our example, we will assume a wire diameter of .62 inches.

Determining the number of active coils needs some judgment on your part. The key work here is active. If the ends of the spring are flattened or ground, these coils are not active, since they are resting on their mounting pads. When counting acting coils, only consider those that can move as the spring is compressed. For the example, assume a spring with 10 active coils.

Determining the mean coil diameter can be done with a tape measure, from the end of the spring, as shown in Figure 5-3. For the example, assume a mean coil diameter of 5.00 inches. Filling in the formula with measured numbers will result in:

$$\text{Coil Spring Rate} = \frac{11{,}250{,}000 \times .62 \times .62 \times .62 \times .62}{8 \times 10 \times 5.0 \times 5.0 \times 5.0}$$

$$\text{Coil Spring Rate} = \frac{1{,}662{,}337}{10{,}000} = 166 \text{ lbs-in.}$$

Several things should be noted from this calculation.

1. If we cut one coil from the spring it will have less active coils, its length will change and the rate will go up:

$$\text{Coil Spring Rate} = \frac{11{,}250{,}000 \times .62 \times .62 \times .62 \times .62}{8 \times 9 \times 5.0 \times 5.0 \times 5.0}$$

$$\text{Coil Spring Rate} = \frac{1{,}662{,}337}{9{,}000} = 187 \text{ lbs-in.}$$

This is about an 11% increase or proportionate to the amount cut from the spring.

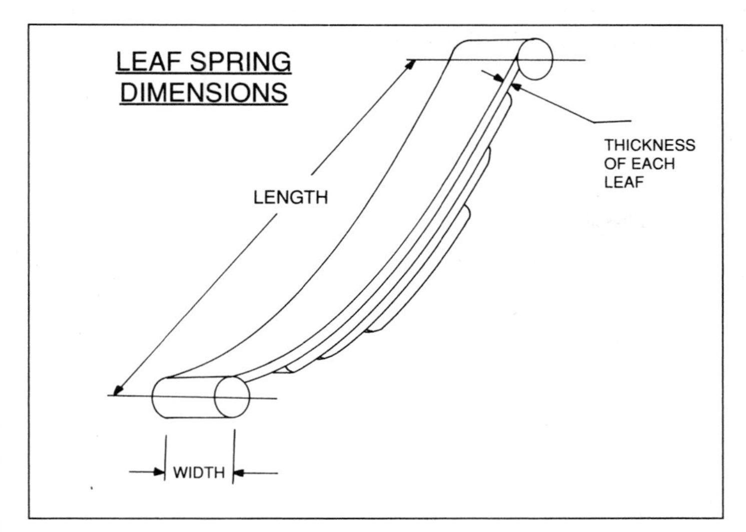

Figure 5-4. When calculating leaf spring rate, it is necessary to know the dimensions of the spring. This diagram shows where the measurements must be made to do these calculations.

2. Increasing the wire diameter will cause a dramatic increase in the rate of the spring. If we increased the wire diameter only 1/32 of an inch the rate would be:

$$\text{Coil Spring Rate} = \frac{11{,}250{,}000 \times .65 \times .65 \times .65 \times .65}{8 \times 10 \times 5.0 \times 5.0 \times 5.0}$$

$$\text{Coil Spring Rate} = \frac{2{,}231{,}328}{10{,}000}$$

Coil Spring Rate = 223 lbs-in.

3. Nothing in the spring rate calculation indicates that a coil spring ever has a change in its rate. The rate is determined by the material, steel in this case, and the dimensions of the spring. Coil springs don't wear out or lose their rate.

4. Nothing in the spring rate calculation describes the spring load. Spring load determines how much weight a spring can support at a given height. The spring rate only tells how much the height will change as the load is changed. A spring can lose its load rating over time or if the steel is not heat-treated properly. However, when a spring sags, its rate is still the same as when it was new.

Calculating Leaf Spring Rates

Leaf spring rate is determined by the spring's physical dimensions as shown in Figure 5-4. The spring rate of a leaf spring can be approximated with this formula:

$$\text{Leaf Spring Rate} = \left(\frac{WN}{12}\right) \times \left(\frac{1000t}{L}\right)^3$$

W = Width of Leaves (inches)

N = Number of Leaves

T = Thickness of 1 Leaf (in inches)

L = Length of Spring (in inches)

12 = A Constant for all Leaf Springs

For the example, assume the following specifications:

W = 3.00 in.

N = 5.0

T = .25 in.

L = 60 in.

Plugging these numbers into the formula results in:

$$\text{Leaf Spring Rate} = \frac{3.00 \times 5}{12} \times \left(\frac{1000 \times .25}{60}\right)^3$$

$$\text{Leaf Spring Rate} = 1.25 \times (4.17)^3$$

Leaf Spring Rate = 90 lbs-in.

WHERE TO MOUNT SPRINGS

Where the springs are mounted is as important as their rate. Either an existing car or a newly designed car must take these factors into consideration.

Front Springs

To minimize the linkage ratio effects, it is usually best to mount the front springs as close to the ball joint as possible. There are obvious clearance considerations which must be resolved, but in general, the closer the better. On Winston Cup cars, which still use 5.0 diameter springs, it is impossible to get the springs very close to the ball joint. This results in big linkage ratios and the need for very high spring rates.

It is also a good policy to mount the springs in a nearly vertical position to keep the spring force at its full value.

Rear Coil Springs

It is often possible to mount rear coil springs directly on the axle housing. This configuration produces wheel rates equal to the spring rates. The rear springs can also be mounted on the rear control arms, but since this introduces linkage ratio considerations, the wheel rate will not be the same as the spring rate.

When mounting the rear springs, it is also necessary to consider the angle of the springs from vertical as this angle can affect the actual spring force. Angling the springs from vertical can reduce the effective force the spring can provide in the vertical direction. A slight angle has little effect, but anything over 30 degrees will result in a significant reduction in the vertical force on the axle, which reduces the spring's effectiveness.

Rear Leaf Springs

Since rear leaf springs are also used to transmit acceleration and braking forces, their positioning and configuration is more critical. These forces cause the axle housing to twist, and in doing this, the leaf springs tend to wrap-up and distort. To counteract this condition, it is necessary to increase the number of leaves, or the thickness of the leaves, to give the spring enough stiffness to resist these forces. Unfortunately, adding leaves or making them thicker, also increases the spring rate. This conflict is solved by adding leaves only to the front of the leaf spring, which provides the spring with the stiffness in the front half it needs to resist the axle torque reaction while not increasing the spring rate as much as if a full leaf was added. Most leaf springs used on high performance cars have extra leaves only on the front of the spring for this reason.

JOUNCE BUMPERS

Jounce bumpers are rubber blocks that keep the lower control arms from hitting the frame on severe bumps. Many enthusiasts throw these away which is a mistake, because they don't realize that these bumpers are really variable-rate springs. By using jounce bumpers, it is possible to use softer spring rates. In the free travel before the jounce bumpers touch the frame, only the car springs are active in

Winston Cup cars use 5.0-inch diameter springs, so it is impossible to get the springs very close to the ball joint. This results in big linkage ratios, and the need for very high spring rates.

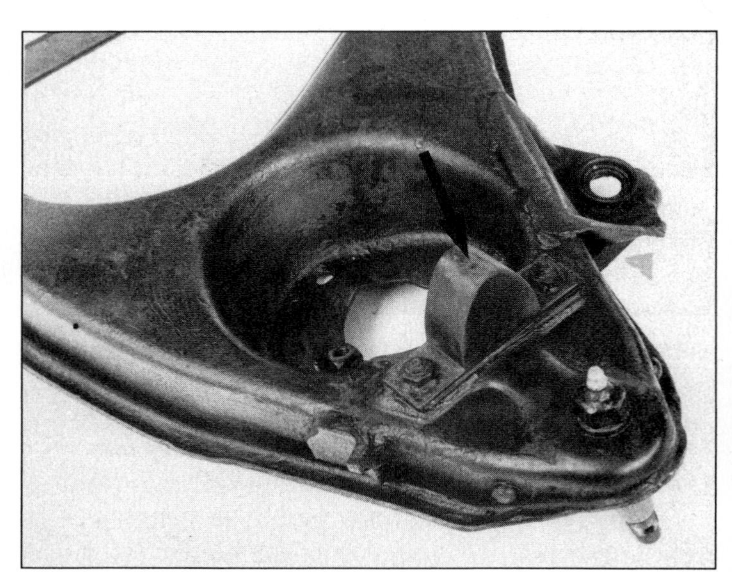

This is a typical rubber jounce bumper installed on a lower control arm. The jounce bumper prevents the lower control arm from hitting the frame on severe bumps. Don't throw them away, like some enthusiasts do—they act as variable-rate springs.

the suspension. On severe bumps, the jounce bumpers are compressed between the frame and the control arms, and the effective spring rate of the suspension is increased. Jounce bumpers are shaped to provide an ever-increasing rate as they are compressed. It is a mistake to trim the shape of a jounce bumper because it destroys the bumper's progressive rate characteristics. Jounce bumpers can be used on race cars for the same reasons they work on production cars (Figure 5-5).

SPRING SELECTION

All of the above is good background information, but in the end you have to make the decision on which springs to use on your car. Understanding all of the factors involved will help make that decision, as well as help to rationalize the springs you are already using. The following are some of the factors that should be considered in choosing springs for your car.

Aerodynamic Effects

Some race cars, like Indy racers, can have aerodynamic downforce that is greater than the weight of the car. If you have a 2000-lb. car with 2000-lbs. of downforce, you will need spring rates and spring loads twice as high as if there were no aerodynamic downforce. A stock body car on a 1/4 mile track does not have much aerodynamic downforce, but some Modified cars do have a considerable amount.

Spring Rates Left-to-Right

Circle track cars will have a large percentage of their weight transferred to the right side of the car during cornering. Since the effective weight on the right side is increased, it follows that more spring rate and spring load will be required. Keeping softer spring rates on the left side is a popular choice, providing it doesn't upset the overall balance of the car. Road-racing and street-driven cars usually use the same spring rates and spring loads from side-to-side because they run around both left- and right-hand turns.

Effects of Too Much Spring Rate

Since most race cars can adjust the spring load to set the correct ride height, choosing the best spring rate is usually of the most concern to racers. We have discussed reasons why more spring rate might be desired, but how do you know if you have too high a spring rate? In general, it is best to run as soft

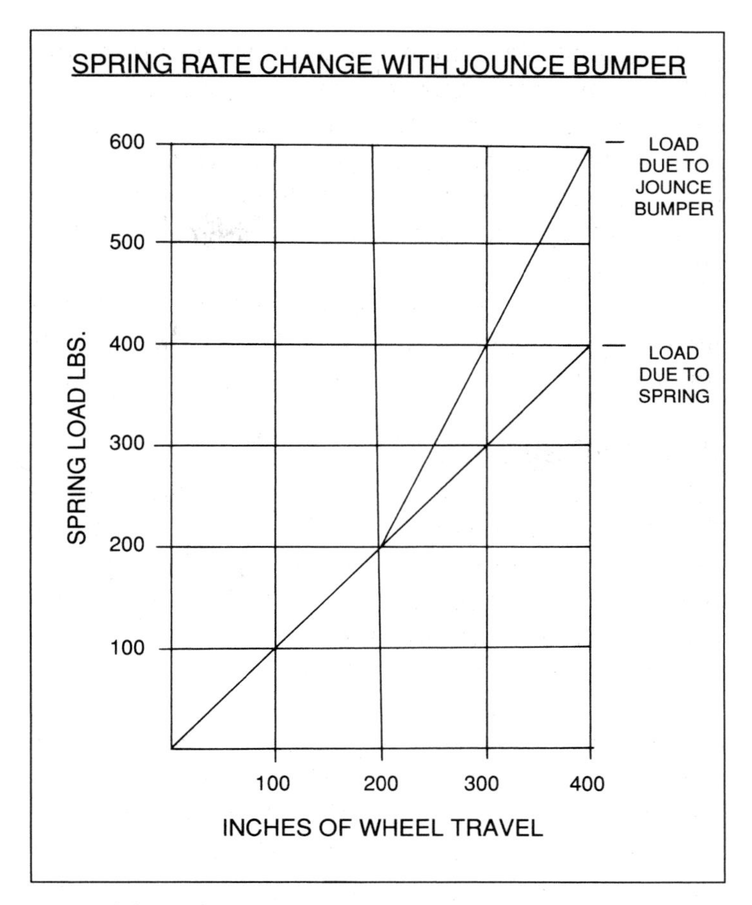

Figure 5-5. When a jounce bumper is used to supplement the regular spring rate, it provides a variable spring rate. The increase in rate helps to prevent the suspension from bottoming out over severe bumps while allowing the use of softer springs necessary for a smooth ride and good cornering force.

a spring rate as possible. The traction between a car's tires and the road is the only source for developing cornering power. If the tires are not in contact with the road, they develop zero cornering power. Soft spring rates allow the tires to better follow road bumps and irregularities so the tires stay in contact with the road a higher percentage of the time. Higher spring rates can also limit suspension in both the bump and rebound directions. If your car is not getting full suspension travel, it might not be allowing the tires to follow the road as much as possible.

Variable Rate Springs—A study of the optimum spring rates for a car eventually come to the conclusion that the best spring rate is a variable one. We would like to have a soft spring rate to absorb the road irregularities and then have a high spring rate to absorb the larger bumps. Springs are available that are made with a variable rate. You can identify these coil springs by noticing that the spacing between the coils is different from one end to the

Indy cars generate tremendous aerodynamic downforce, especially on high speed ovals like Michigan and Indianapolis. Springs with more rate and load are needed.

other. The design concept for these springs is correct, but in practice they don't work too well. Since car springs only operate in a limited percentage of their total travel, the amount of change in rate is usually pretty small. Jounce bumpers provide this same variable rate feature with more consistency.

Spring Rates for Oval Tracks

When a car runs on a banked track, spring rates must be increased to compensate for the increase in effective weight of the car. Chart 5-2 gives an approximation for how much spring rates must be increased for various bank angles. These calculations are based on the car maintaining a 1.00 cornering force at all bank angles. Since a car might not be able to maintain such a cornering force because of aerodynamics or horsepower, these figures might be high under some circumstances. These values are close, but the car and the driver should be the final element in choosing spring rates at any given track.

Spring Rates for the Street

The purpose of suspension springs is to hold the car steady while allowing the wheels to follow road irregularities. In general, the softest possible springs will do this job best. Softer springs will allow each individual wheel to move in relation to the chassis while having the minimum effect on the driver's compartment. This translates into a soft ride, noise isolation and good handling.

All stiffer springs do is make the car have a stiff ride. They have no capability to make a significant improvement in handling. As long as the springs on a car are stiff enough to keep the car from bottoming out, they are adequate. If a car is lowered, a slight increase in spring rate can be used to compensate for the reduced ride travel.

Some car enthusiasts mistakenly believe that if 300 lbs-in. coil springs are good, then 600 lbs-in. springs have to be better. They're wrong. Optimum road-holding demands that the tires be in contact with the pavement; a soft spring lets the wheels follow road irregularities so that the tires can generate maximum adhesion.

Our recommendation for front springs on a street-driven car is to use the standard factory coils. For street use, you can trim one half coil off the top of

Chart 5-2

Bank Angle of Track (in degrees)	% Increase in Spring Rate
0	0%
5	10%
10	23%
15	41%
20	67%
25	106%

the spring with an acetylene torch to lower the car slightly (see sidebar p. 35).

Most front spring rates range between 300 and 350 lbs-in. Trimming the coils as we've recommended will increase the rate approximately 10%. But the true purpose of trimming the front springs is to lower the car for improved aerodynamics and better handling, not to increase the spring rate.

Several cars, like the Corvette with F-41 suspension, the WS-6 Trans Am and the later model Camaro Z-28's, are equipped with very high rate springs. We have found that the ride can be greatly improved by installing softer springs, with no effect on the handling capabilities of these cars. Many enthusiasts buy the standard rate soft springs from their dealer for these cars and cut 1/2 coil for the proper ride height.

ADJUSTING RIDE HEIGHT

Although the spring rate will not change during the life of a spring, the spring load can change. This is commonly called *spring sag*. Loss of load, or spring sag, can be caused by a variety of reasons, including poor metallurgy, overloading and even fatigue due to high mileage. If a spring has lost load, this problem should not be corrected by changing the spring rate. If a spring has sagged, you can get the spring load back to normal using several methods. For coil springs, the usual procedure is to place a rubber shim on top of the spring to increase the load. (These spring shims are available through car dealership parts departments and from auto parts stores.) A leaf spring's load height can be increased by using different shackle lengths or, in extreme cases, by re-arching the spring.

Lowering—Before you invest in springs to lower your car, you should be aware that your chances of success are much better if you simply cut your existing springs. All springs take a certain amount of permanent set after they are installed. (Note that new cars sitting in the showroom are higher than the same models with only a few thousand miles on them.) This change in height is caused by the inevitable loss in load that any new spring experiences. If you put new springs on your car, they will settle as much as an inch within the first few months. Obviously, if your car was the correct height when you installed the new springs, it will be too low after a few months.

This is why I recommend you cut your existing springs and save the cost of new springs. The results will almost certainly be better. Your car's existing springs have already taken a permanent set, so you know where they will end up. If you want to lower your car, you can cut your existing springs to

Unlike circle track cars, which use more spring rate and load on the right side of the car, road-racing and street cars usually use the same spring rates and loads on both sides, because they will be turning right and left. Photo by Michael Lutfy

achieve the ride height you want with one operation. See the sidebar nearby for details.

Springs don't wear out, so you can save the cost of new springs if you trim your existing ones. If you want to feel like you bought new springs, paint your old ones. We have found that most coil springs should be cut 1/2 coil to lower the car and still keep adequate ride height. If your springs have sagged, you don't even have to cut them because the car is already at a lower ride height.

SHOCK ABSORBERS

The purpose of shock absorbers is to control the velocity of the suspension. If the shocks don't have enough resistance, the spring will move the suspension too fast and it will have an underdampened motion. If the shocks are too firm, the motion will be overdampened. It is important to have just the right amount of dampening to control the spring action of the suspension.

Proper Dampening—Extra firm shocks have the same negative effects on ride and handling as extra stiff springs. The tires cannot follow the road irregularities unless they are free to move in relation to the chassis. The relative motion must be as free as possible, but it also must be controlled. Shocks are used for control. If the shocks are too firm, the suspension will be over-dampened and the tires will not be able to keep in contact with the road. If there is no dampening, the suspension will cycle up and down at its natural frequency and again the tires will not keep in contact with the road. What is needed is *critical-dampening*, which means just enough control to keep the suspension from cycling. We have found that adjustable shocks allow each car to be tuned for critical dampening. Run your shocks as soft as possible—just enough so the car doesn't wallow over bumps (Figure 5-6).

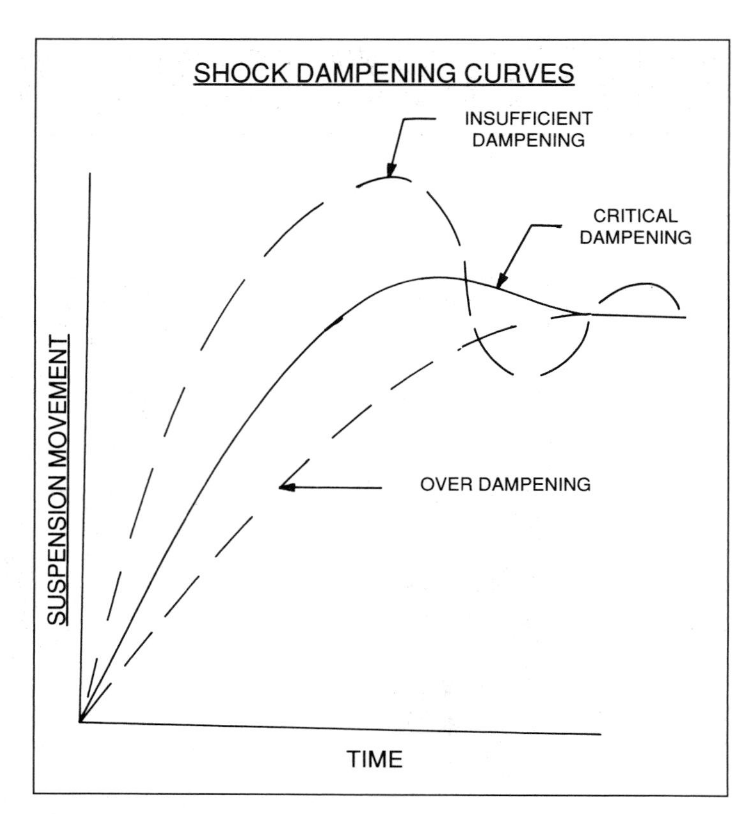

Figure 5-6. In order to provide the correct amount of dampening for the spring in a suspension system, it is necessary to have the critical amount. Overdampening will restrict the free movement of the suspension and insufficient dampening will allow the suspension to cycle freely without control.

Mounting Shocks

A shock absorber's ability to control the suspension is dependent on how it is mounted. For street-driven cars, rubber-insulated mounting points are preferred because they help to dampen road noise. For race cars, rod-end type mounts are preferred because they prevent the suspension from moving without moving the shock piston.

Front Shocks—The same consideration for mounting the front springs should be used for mounting the front shocks. Basically, this means placing them as close to the ball joint as possible. This location increases their effectiveness. You can mount them further from the ball joint and then increase their dampening rate to get the same control. This procedure, however, magnifies any clearances or loose-

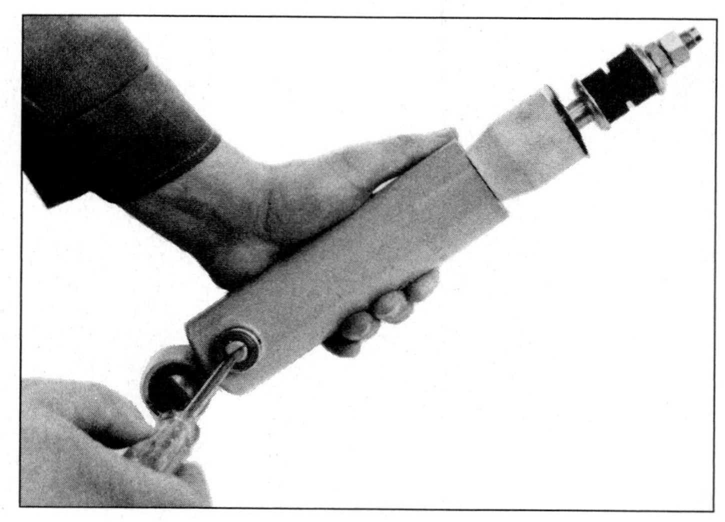

Most production shocks come with rubber insulators to provide a better ride and lower suspension noise. However, during high performance use, these rubber components can deflect.

CUTTING COIL SPRINGS

1. Carefully remove the coil springs following the procedure outlined in your car's shop manual. Note how the end coils of this stock front spring are bent slightly so that it seats properly in the frame and A-arm.

2. Most stock springs can be trimmed a half coil to provide the proper ride height for improved handling. Mark the spring directly across from the original end as shown.

3. Cut the spring with an acetylene torch.

4. Heat the half coil below your cut so you can bend it to match the spring's original shape.

5. Quickly turn the spring upside down and bend the top coil by pushing down on the spring. DO NOT quench the spring with water; allow it to air-cool slowly.

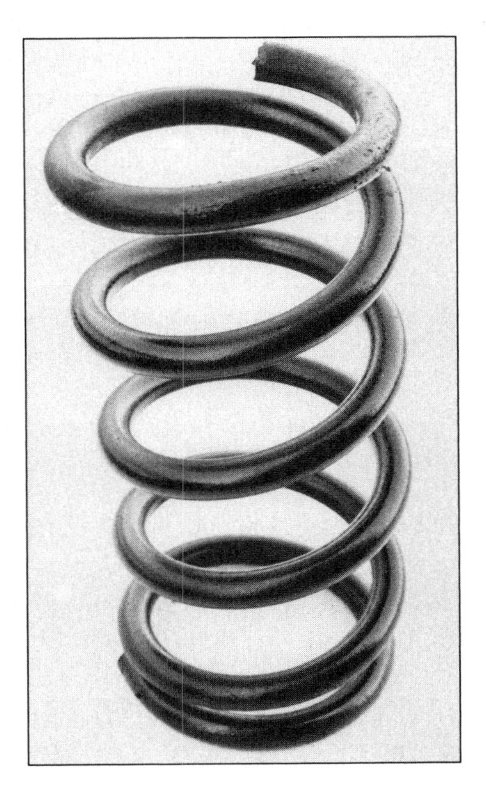

6. Paint the spring and reinstall it according to the directions in your shop manual. Realign the front suspension.

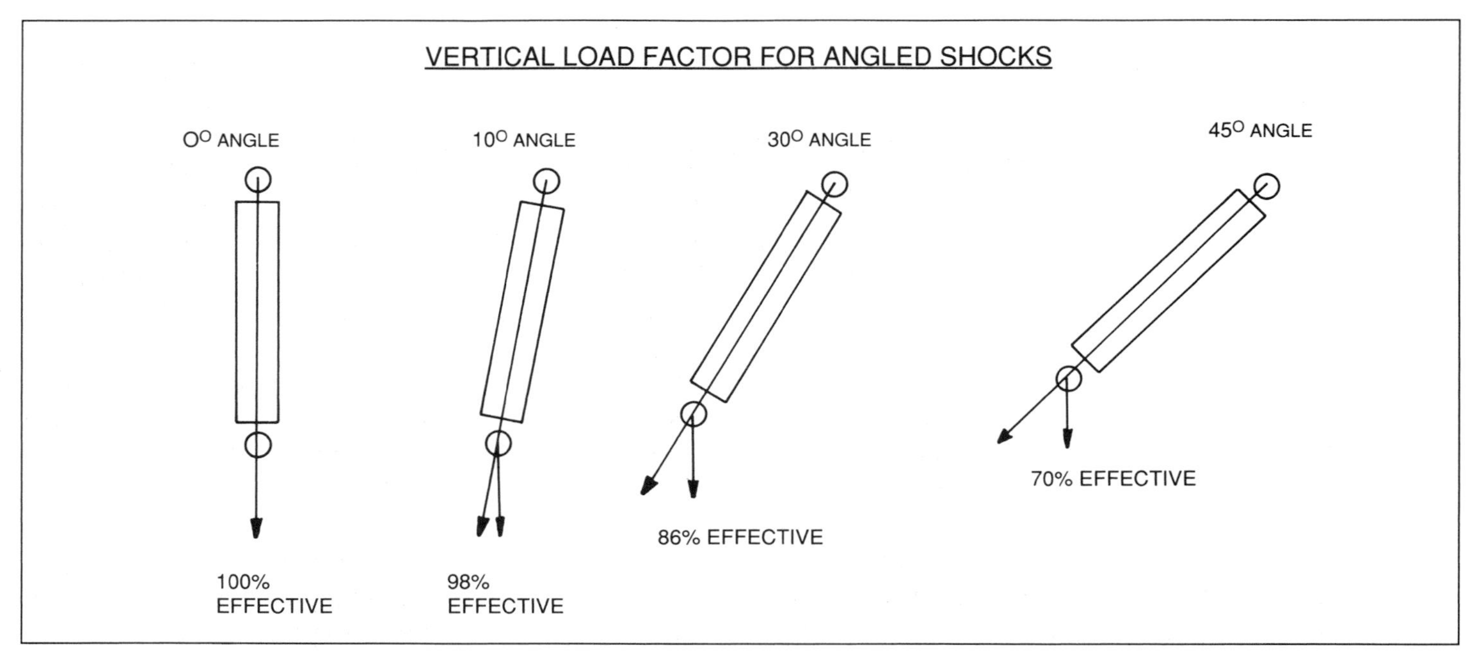

Figure 5-7. Shocks are most effective when they are mounted perpendicular to the direction of travel. Slight variations are permissible, but anything over 30 degrees will result in a loss of shock effectiveness.

ness in the system which can result in loss of control at the wheel.

Rear Shocks w/Coil Springs—All those considera-tion mentioned above for spring and shock location should be followed in mounting rear shocks on a coil spring type rear suspension. Basically, keep the shocks as close to the wheel as possible and prevent any lost motion between the shock and the axle. Mount the shocks as nearly vertical as possible to maximize their effectiveness.

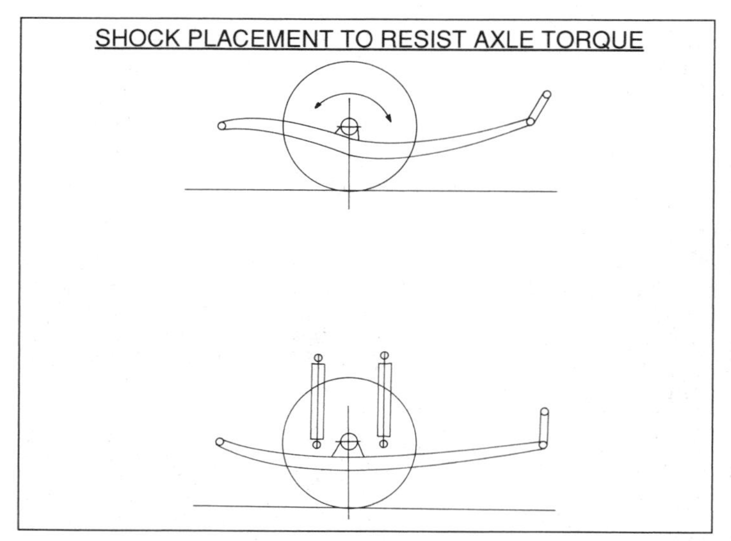

Figure 5-8. On cars with a Hotchkiss drive rear suspension, the shocks can be placed on either side of the axle housing to help prevent the axle housing from rotating. When the axle housing is free to rotate, power hop and/or brake hop will usually result.

Rear Shocks w/Leaf Springs—Mounting the shocks on a leaf spring rear suspension is a special case which requires special considerations. Chevrolet found that if one shock is placed in front of the axle, and one shock is placed behind the axle, their damp-ening effects do a good job of controlling the violent torsional rotation of the axle during *power-hop* and *brake-hop*. This configuration is so successful that it eliminates most of these problems without any other devices. Under extreme conditions, traction bars and/or telescoping brake rods can be used to provide additional torsional control of the axle (Figure 5-8). For more information, turn to Chapter 9, page 72. ■

Because of the deflection, the shocks are prevented from acting immediately on the suspension, so they are usually replaced with rod-end type mounts on race cars.

TYPES OF FRONT SUSPENSION 6

The evolution of high performance front suspension design has been ongoing for nearly 100 years. During that time, there have been many ideas presented in an effort to find the best performance configuration. The criteria for determining the best configuration is very simple, "What makes the car get around the corner the quickest?" Optimizing the design aspects of each configuration is a more complex problem. The technology of high performance front suspension design has been closely related to the technology of production-car front suspension design, so the two must be studied together.

The factors that are used to determine a good production car front suspension design, however, are not always the same as those that are used to determine a good high performance front suspension design. In the 1930's, production cars started to use independent front suspensions, but it wasn't until the 1960's that it was developed to provide good cornering power.

Analyzing the types of front suspension designs that have been used in the past will help you to understand how the current systems evolved.

BEAM AXLE

A beam axle is one where both wheels are connected to a rigid axle. Early cars used a beam-axle front suspension as a carryover from horse-drawn carriages, and because they worked well, there was no need to consider other types. Even today, beam-

The new-for-'93 Pontiac Trans Am utilizes an unequal length double A-arm front suspension, what they call an "SLA" (Short and Long Arm) front suspension. This type of front suspension causes the tires and wheels to gain negative camber as the suspension compresses, which means the camber angle between the outside tire and the ground will not change much in relation to the ground during body roll. This allows the outside tire to develop maximum cornering power.

axle front suspensions are used on most sprint cars and midgets. Indycars used beam-axle front suspensions up until the 1960's, and they are still used on tractor-trailer trucks and some 4WD trucks. Beam-axle front suspensions are used in these applications because they provide a good solution to the problem of suspending the vehicle from the front wheels. Beam axles are still used on the rear of many cars, so this "old" technology still has a useful purpose today. Some new technology was applied to the beam axle when coil springs and torsion bars were used to replace the leaf springs. These improvements in springing, and similar improvements in shock absorbers, have kept the ancient beam-axle front suspensions competitive with the newer independent types for some applications.

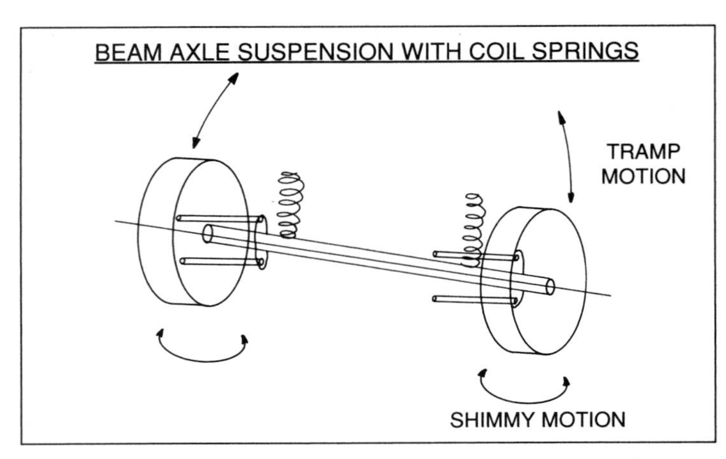

Figure 6-2. One of the disadvantages of a beam axle front suspension is that the gyroscopic forces from one wheel are fed through the axle and the steering linkage to the other wheel. This connection results in shimmy and tramp motions which are difficult to control.

Advantages

The advantages of a beam-axle suspension are:

Strength & Rigidity—When weight is not a big factor, a beam-axle front suspension can be made to be very strong. This factor, and its basic simplicity, are the reasons why beam-axle front suspensions are used on semi-trucks.

Camber Control—Since a beam-axle connects both front wheels together, the camber angle between the tires and the road will always remain constant. Decamber can still be used to advantage, but there will be no loss of camber because of suspension movement or body roll.

Disadvantages

The disadvantages of beam-axle suspension are:

Heavy Unsprung Weight—Because both front wheels are attached to the same axle, its weight affects the unsprung weight of each wheel. This inertia, and the interaction between the wheels, reduces road-holding on anything but a smooth surface. Independent front suspension is considered better than a beam-axle front suspension, because the forces affecting one wheel do not affect the other wheel.

Shimmy, Tramp & Bump Steer—When both front wheels are connected to the same axle beam, the steering forces between the two front wheels are also interconnected. This interconnection can cause the front wheels to *shimmy* and *tramp* under certain conditions. Shimmy and tramp are uncontrolled vibrations of the wheels due to gyroscopic forces reacting between the wheels. The problem can usually be controlled with tight bushings and joints, and with effective shock absorbers, but it is a fundamental disadvantage with a beam-axle front suspension. Because of the axle control devices used to position a beam-axle front suspension, it is difficult to locate the steering linkage so that the

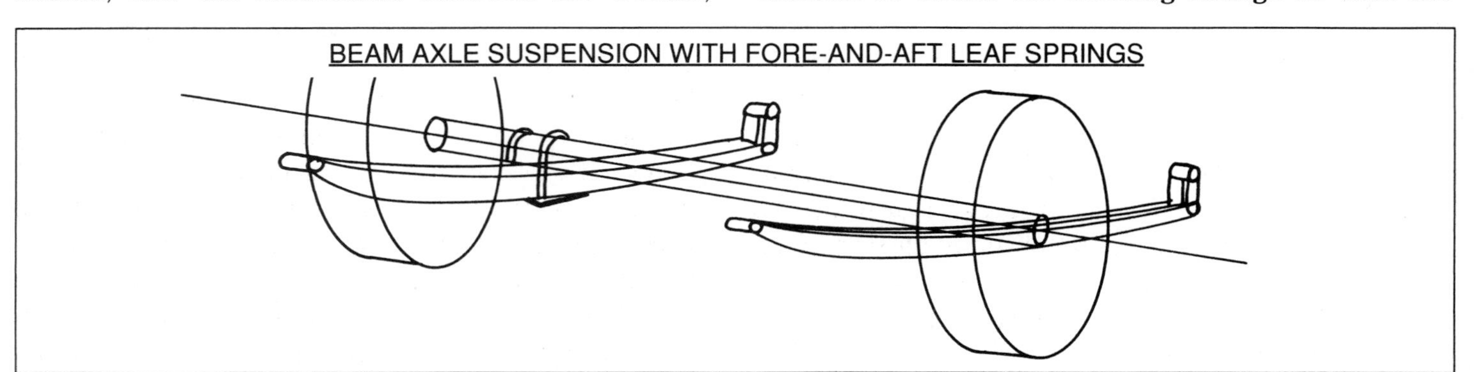

Figure 6-1. The beam axle front suspension is the most basic type, in that both wheels are attached to the same axle. This is a very rugged arrangement, but it has several disadvantages, such as heavy unsprung weight, shimmy, tramp and bump steer, that make it less than desirable for high performance cars. Shown here is an axle located with fore-and-aft leaf springs like those used on many trucks.

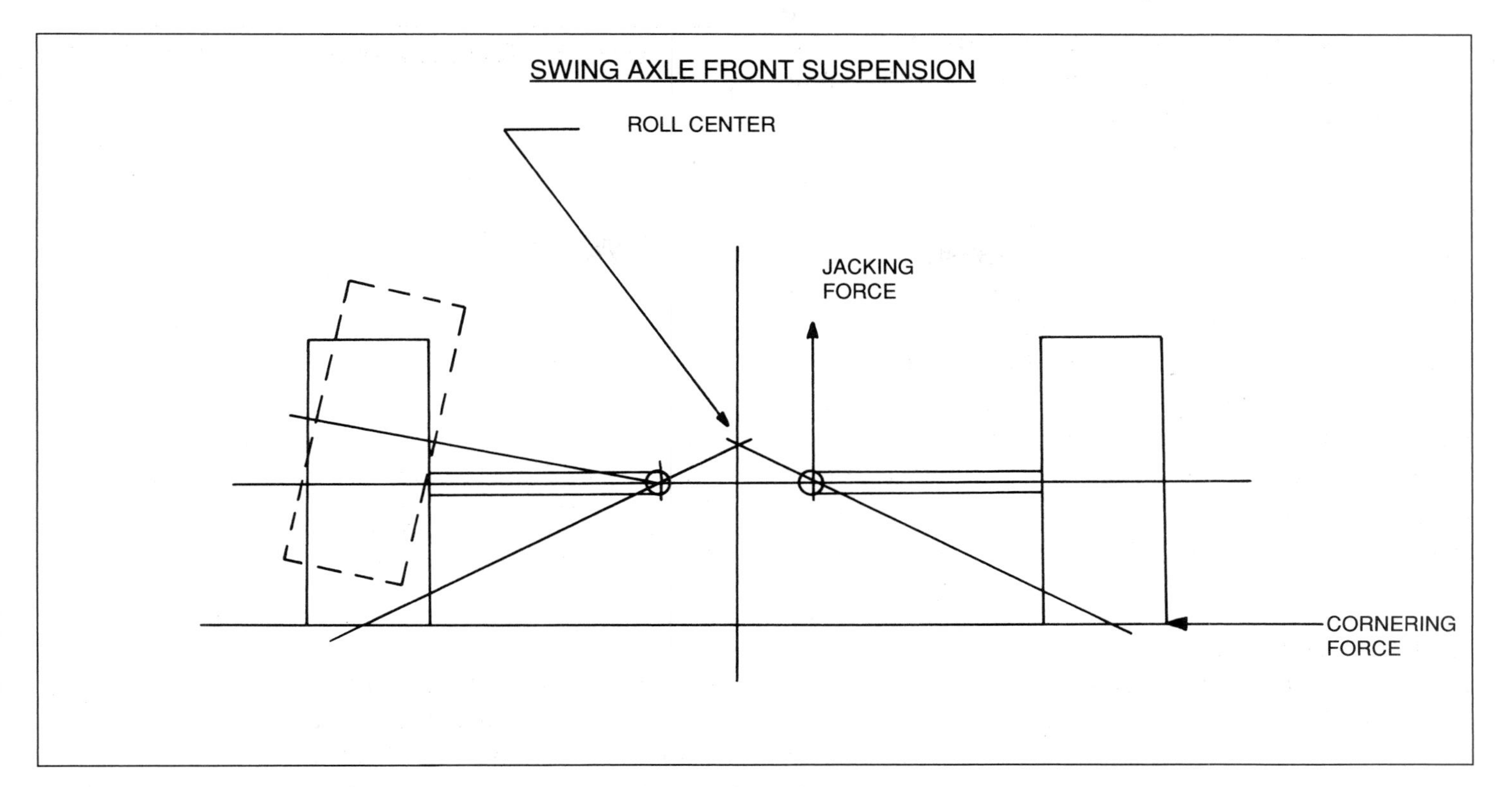

Figure 6-3. A swing axle front suspension has a high roll center and a short swing arm length, which looks good an paper because it provides a lot of camber gain. However, in practice, cornering forces react through the axle shaft, which jacks up the car. When the car is jacked up, it causes positive camber on the outside wheel, which results in a severe loss in cornering power when it is needed most.

bump steer is minimized. Various methods have been used to solve this problem, but again it remains a fundamental disadvantage with this type of front suspension.

Space Requirements—A beam axle with the related suspension arms and steering linkage requires a considerable amount of space if the suspension travel is to be enough to provide good ride motions. This disadvantage of a beam-axle front suspension was one of the major reasons why production cars started to use independent front suspension.

Rough Ride—The major reason why production cars started using independent front suspension was because it improved the ride characteristics of the car. The heavy unsprung weight of the beam-axle suspension, and the interconnection of the front wheels, does not allow for as much ride isolation between the wheels and the chassis, so a rougher ride results. Ride quality is also important for road-holding since those factors that improve the ride characteristics also help to improve the road-holding, because they keep the tires in better contact with the road.

SWING AXLE

A swing-axle suspension system has a pivot near the center of the car which allows the axle to swing (Figure 6-3). One of the first attempts to provide an independent front suspension, and eventually an independent rear suspension, was with the swing axle. This is a simple and strong system that offers all of the advantages of an independent suspension, whether it is used on the front or on the rear. All of the early Volkswagen Beetles used a swing axle on the rear, and some Ford pickup trucks still use it on the front.

Disadvantages

Although the design of a swing-axle suspension is satisfactory, development and testing have shown there are some serious handling problems with its use on vehicles that can generate high cornering forces. The basic problem is called *jacking* and it results from the cornering forces acting through the wheel and axle to raise the car. When the car is jacked up, camber angle is reduced and some of the cornering power is lost. Since the jacking only happens during hard cornering, the loss of cornering power happens when it's needed most. And, since the jacking happens quickly, the resulting change in handling is sudden and severe. The early Corvairs made this problem famous because they would suddenly spin-out if the driver attempted to corner at the limits of the car's traction.

Another way to analyze a swing-axle suspension is to notice that its roll center is high and its swing arm length is short. Since the combination of a high roll center and a short swing arm length can cause erratic handling on a swing-axle suspension, it is not wise to use this same combination with other types of suspension.

TRAILING LINK

A trailing-link suspension uses two arms to support the steering knuckle which trail from ahead of the knuckle (Figure 6-4). Since independent front suspensions for production cars were first developed to improve ride quality and to reduce the space required for the suspension, early designs tried to maximize these advantages. One of the more popular configurations was the trailing-link type of front suspension, like the one used on the Volkswagen Beetles. This design uses a minimum of space and offers the possible advantage of dragging the wheels over bumps, so that the suspension not only raises, but it also moves backwards. From a ride standpoint, the trailing-link front suspension was a success. It offered a ride quality much better than thought possible with a beam-axle, so the production car manufacturers liked it.

Disadvantages

The trailing-link front suspension design was not adapted to many high performance cars because there are serious design problems that show up during hard cornering. When cornering loads are applied to the trailing links, they bend. Road irregularities cause various amounts of bending and the result is severe vibration. This vibration shakes the tires and wheels so the steering vibrates and the tires don't adhere to the ground.

Another problem with a trailing-link front suspension is that the camber angle of the wheels changes in relation to the ground when the car rolls during cornering. This loss of camber angle severely reduces the cornering traction of the front tires, so cornering speeds are reduced. With a trailing-link front suspension, the swing arm length is infinitely long and the roll center is at ground level. These factors promote body roll and camber loss.

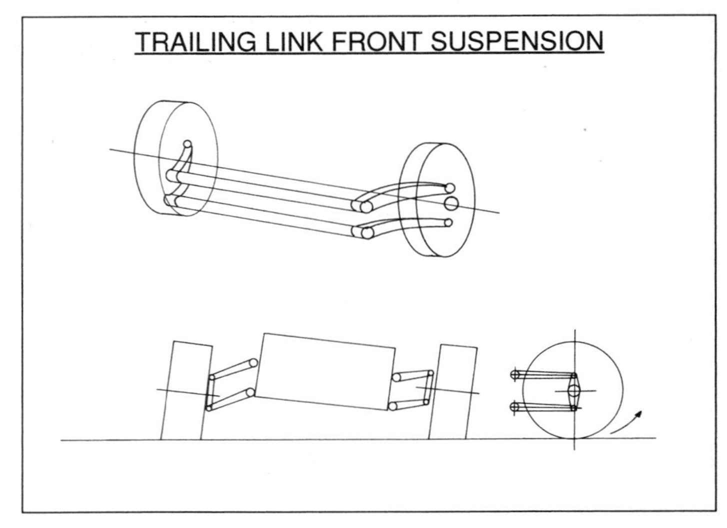

Figure 6-4. A trailing link front suspension has the apparent design advantage of improving the ride by pulling the wheels over the bumps. This feature is negated on the overall picture by the problem of camber loss during body roll, and by bending of the control arms during hard cornering.

Another problem with trailing-link front suspensions is that the control arms need to be heavy in order to absorb the bending forces. The arms also require considerable frame structure ahead of the front wheels, which increases total car weight.

With these problems it is easy to see why trailing-link front suspensions never became popular on high performance cars. These same reasons are why this design of front suspension is also not seen on any current production car.

STRUT

The knuckle on a strut-type suspension is mounted on the shock absorber. Sometimes they are referred to as MacPherson struts (Figure 6-5). Strut-type front suspensions became popular on production cars in the '70's because they offered a simple and inexpensive configuration that doesn't take up much space. It is particularly well-suited to front-wheel-drive production cars, because it allows room for the front drive-axle to pass through the front hub. Most of today's small cars use this type of front suspension, because it is inexpensive and gives a fairly good ride quality with the compact dimensions needed for front-wheel-drive cars.

Disadvantages

Except where it might be required by the race-sanctioning bodies in production-based classes,

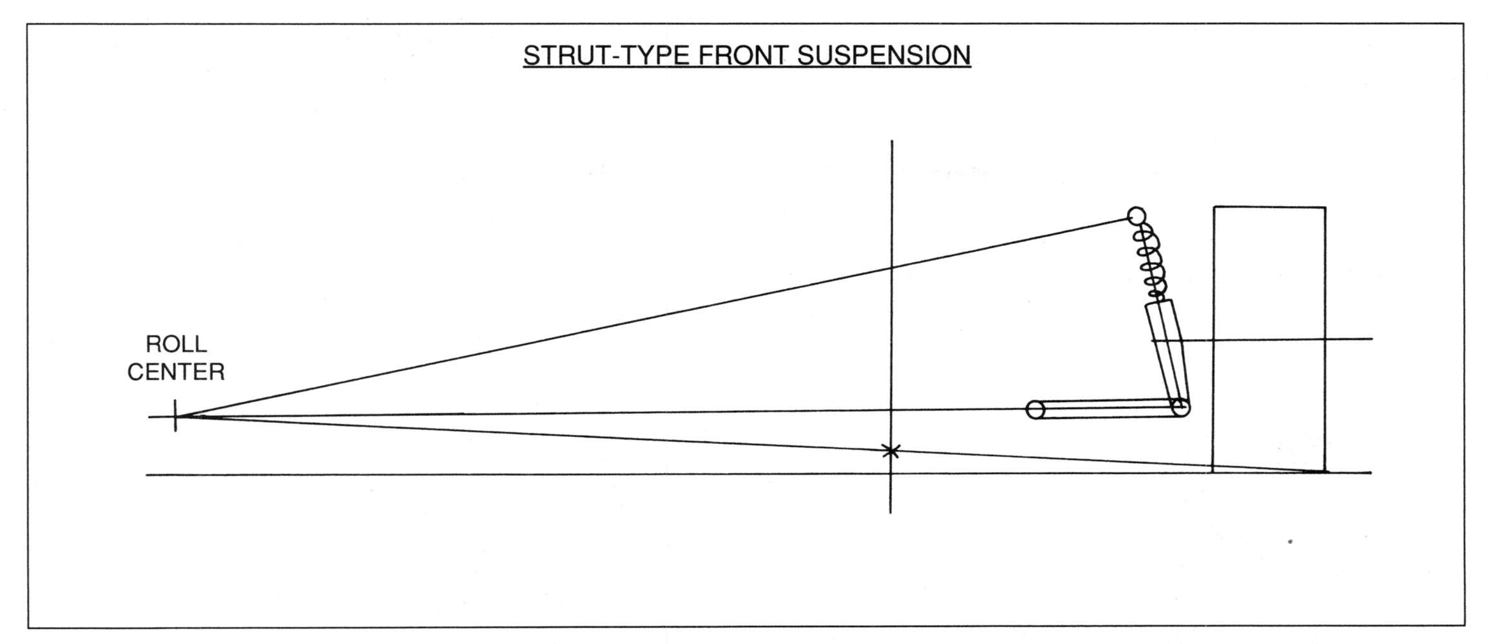

Figure 6-5. Strut-type front suspensions are used on front-wheel-drive cars because they provide room for the drive axle. The problem with strut-type front suspensions is the need for a high cowl height, lack of room for large tires and limited camber gain.

such as Showroom Stock, it probably won't be used on many race cars. One of the problems with a strut-type front suspension on a race car is that there isn't much room for wide tires and wheels without increasing the scrub radius. Increasing the scrub radius causes a big increase in the side loading of the sliding members, which causes bending and higher friction loads. Another problem that con cerns racers is that there is little camber gain possible with this type of front suspension. This means that the outside tire will lose its camber angle to the road as the car rolls during cornering. With a strut-type front suspension, the swing arm length is reasonable and the roll center can be above the ground. It is interesting to note that even with favorable design objectives for roll center height and swing arm length, the lack of sufficient camber gain keeps this type of front suspension design from being suitable for race cars.

Another problem with a strut-type front suspension is that it requires a high cowl height to provide sufficient room for the tops of the struts. This can be a problem when building a low-profile race car.

Honda has quit using strut-type suspension on its production cars for these same reasons. They apparently feel that it is worth the increased cost to equip their cars with the superior double A-arm type of suspension.

Strut-type front suspension is used on some low-cost cars that offer high performance, but it is done to reduce cost, not for any handling advantages.

EQUAL LENGTH DOUBLE A-ARM

The front knuckle on a double A-arm front suspension is supported by a triangulated control arm at the top and at the bottom (Figure 6-6). Early versions of double A-arm front suspension had equal length arms mounted parallel to the ground. With this arrangement, the swing-arm length is infinitely long and the roll center height is at ground level. Unfortunately for high performance handling, this system does not provide any camber gain, so the front tires lose camber as the body rolls and cornering power is lost.

Advantages

A double A-arm front suspension design offers a big advantage over the other types because it uses rigid control arms to connect the front knuckles to the frame. These arms prevent deflections during hard cornering, which insures that the steering and the wheel alignment stay constant. The ride characteristics of a double A-arm front suspension are excellent, so most manufacturers of quality production cars use this basic system. All that is needed to make double A-arm front suspensions satisfactory

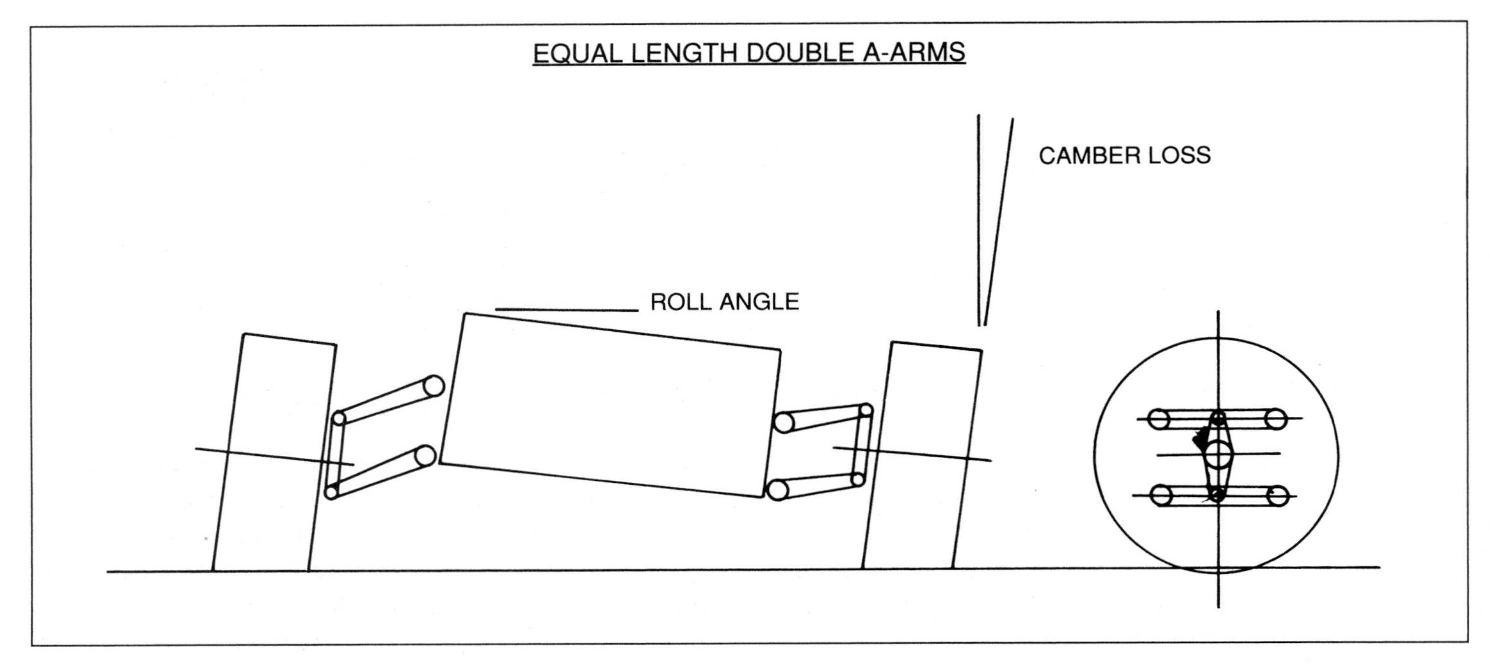

Figure 6-6. Double A-arms are preferred, because they mount the wheel and hub on rigid arms, which do not deflect during hard cornering. The problem of equal length arms is that they do not provide any camber gain.

for high performance use is to find a way to get the system to provide camber gain as the suspension was compressed. This problem was solved by changing the length of the A-arms.

UNEQUAL LENGTH DOUBLE A-ARM

The use of a longer, lower A-arm and a shorter, upper A-arm provide a suspension geometry that causes the tires and wheels to gain negative camber as the suspension compresses (Figure 6-7). The reason this happens is because the shorter upper arm will cause the top of the knuckle to pull inboard faster than the longer lower arm moves the bottom of the knuckle inboard. The advantage of this gain in negative camber is that the camber angle between the outside tire and the ground will not change too much in relation to the ground during body roll. If the outside tire is kept perpendicular to the ground, it will develop its maximum cornering power, and the car will go around turns faster. Since there is little weight on the inside tire, the angle between it and the road is of little concern.

With unequal length control arms mounted at various angles to the ground, it is possible to design for just about any roll center height and swing-arm length. This flexibility in the design process gives a chassis engineer many options on how to lay out a suspension. Computer programs are available for quickly analyzing the variations, so in recent years there has been more refinement in the application of double A-arm suspensions.

How long each A-arm should be, and what angle the A-arms should be to the chassis for optimum geometry, has been studied for over 50 years. Not everyone agrees on what the optimum values are, but Chapter 7 explains some of the considerations involved. There is no right or wrong answer. The bottom line is "whatever makes a car go around corners fastest." Under some conditions, one arrangement might be better than another, and so each situation needs to be evaluated on its own merits.■

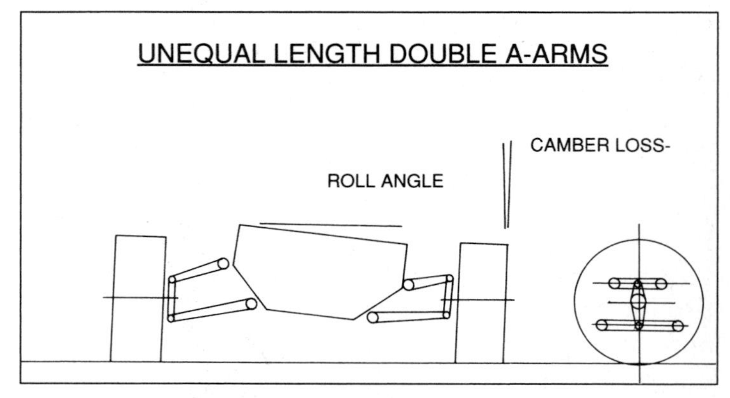

Figure 6-7. Unequal length control arms mounted at various angles can provide a variety of roll center heights and swing-arm lengths. The amount of camber gain is determined by the length of the swing arm and the height of the instant center.

FRONT SUSPENSION DESIGN

The factors involved in front suspension design include the components and their arrangement on the chassis. All of these factors are related, so each element must be optimized on its own merits as well as to how it fits into the system. The following specific design aspects are usually considered in laying out a new front suspension. After the layout is established, it can be analyzed with a computer program to determine how it will move throughout the suspension travel. Unfortunately, the computer analysis won't tell us how well the suspension will work without interpreting the data.

SUSPENSION GEOMETRY

All the factors that are involved in the geometry of an independent suspension system (front or rear), make its study a complex problem. In order to explain each element, and how it fits into the total system, I will go through a design exercise. Hopefully, this step-by-step process will give an overview, as well as consideration for each aspect in its own right.

Before starting a suspension design, it should be understood that there are no absolutes. Each design factor must be evaluated as it fits into the

Front suspension design requires that you consider all components and their arrangement on the chassis. All factors are interrelated, so each element must be optimized on its own merits as well as to how it fits into the overall system. Photo by Michael Lutfy.

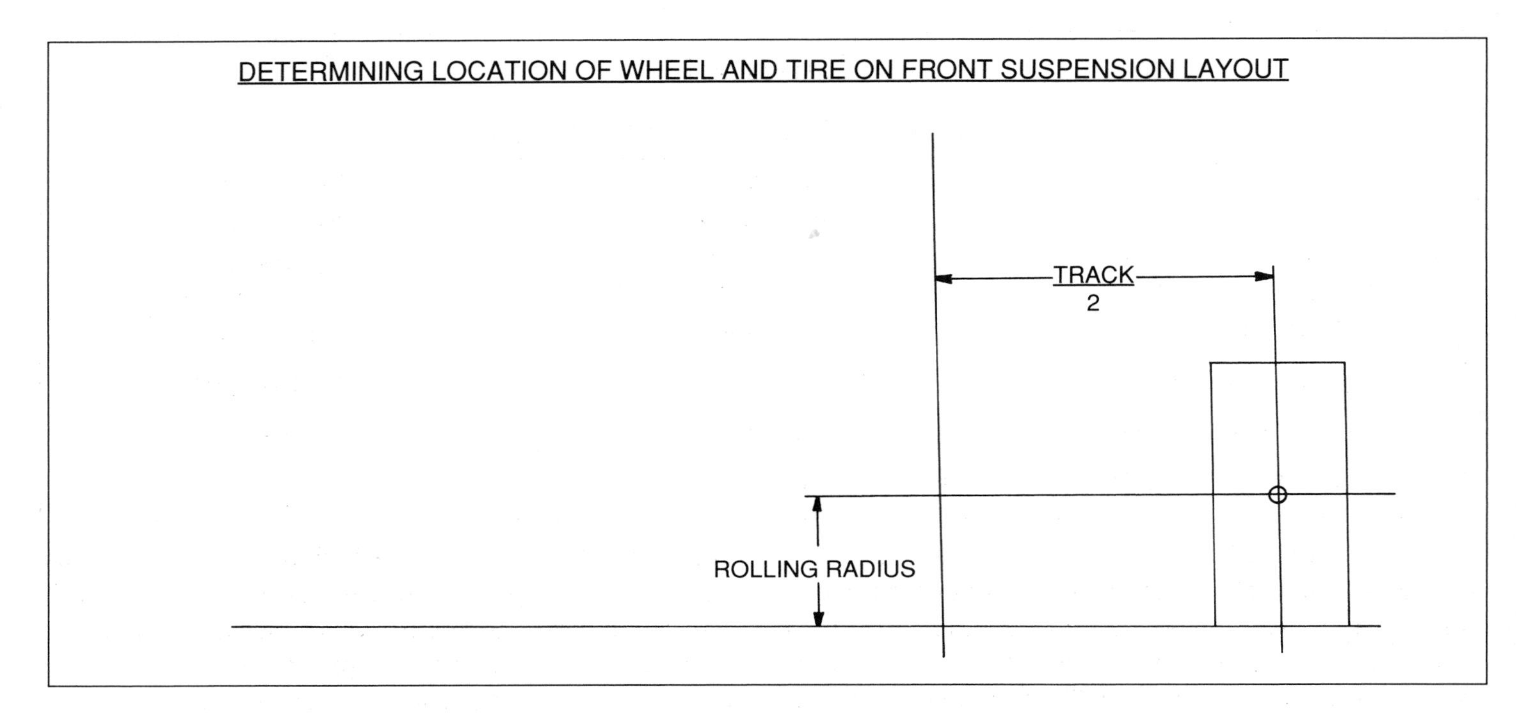

Figure 7-1. A suspension is designed from the wheels inboard, so the first thing to establish is the size and location of the wheels and tires. You can use as much detail as you wish, but we like to keep it to the minimum because of likely changes.

overall system. Every design decision is a compromise based on several other factors. Over the past 30 years, most of the factors relating to independent suspension design have been thoroughly tested and developed. The science has been researched and the results are pretty much agreed upon. Radical departures from proven practices will probably result in an ill-handling car. The values I will present in the following design exercise are ones that have been proven to work correctly. You can use other values on your car, but you should understand how these variations can affect your car's handling.

A car's suspension design starts at the tires and wheels and moves inboard. The last thing that is designed is the frame.

Wheels, Tires & Offset

To start your design layout, you should first draw a front view of your wheel and tire setting on the ground (Figure 7-1). This is a simple matter of measuring the wheel and tire in cross-section and transferring the dimensions and shape to the layout. *Wheel offset*, which is the distance of the wheel rim to the center of the mounting flange, is an important design dimension, so this should also be shown. If larger tires or wheels might be used, you should also show these on your layout. If you have a choice of wheel offset, it is a design advantage to use wheels that permit the knuckle to be placed inside the wheel. Wheels with the offset toward the outside of the car might look trick, but they result in other suspension problems that overpower their good looks. For our example, I am using a 60-inch track and 10x15-in. wheels with a 1.50-in. positive offset. This means that the wheel mounting surface will be 28.50 inches from the centerline of the car. Ideally, we would use a 1.50-in. negative offset, which would make the wheel mounting surface 31.50 inches from the centerline of the car. The following discussions will show why this would be preferred.

Our tire for this design study has a rolling radius of 12.5 inches and a section width of 13.0 inches.

Front Track

The front track is usually determined by car width, so put these dimensions on your layout. Obviously, the centerline of the car will be midway between the tire centerlines, so this dimension can also be put on the layout. If you are not required to run a given track dimension by your racing rule book, you have a choice to make that involves some compromises. A wider track will improve cornering power, because it reduces weight transfer in the corners. Less weight transfer means that the tire loading will be more equal and the cornering power will be greater. Unfortunately, a wider track dimension will make the

Chart 7-1

BALL JOINT PIVOT LOCATIONS

	Height from Axle Center (inches)	Offset from Wheel Surface (inches)
Lower Ball Joint	5.75 below	3.75 inboard
Upper Ball Joint	5.75 above	5.50 inboard
Steering Ball Joint	5.25 below	3.50 inboard

Note: Figure 7-2 shows these dimensions added to the layout.

car wider and its frontal area will increase. A wider car can be more difficult to drive in heavy traffic, and more frontal area causes more aerodynamic drag. If you run at low speed, and have plenty of room for maneuvering, a wider track will be an advantage. As mentioned, our sample layout shows a 60.0-inch track, so the distance from the centerline of the tire to the centerline of the car will be 30.0 inches.

Knuckle Design

The knuckle arrangement has a great deal of influence on the total effectiveness of a suspension design. The knuckle determines the position of the upper ball joint, the lower ball joint, and the steering ball joint in relation to the wheel and tire. For our example I will use a cast stainless steel knuckle to show how these factors interrelate.

Ball Joint Heights—Since all the loads from the wheel and tire must be fed into the chassis and control arms through the upper and lower ball joints, it is an advantage to place the ball joints as far apart as possible. Any given load results in lower forces if the spread between the points is increased. The limitations for increasing the spread between the ball joints is usually clearance between the joints and the wheels and tires. The cast knuckle used in our example has been designed to allow placing the joints inside a 15-in. wheel, so that the scrub radius can be reduced if the proper wheel offset is used. The ball joint pivots on our sample knuckle are located as seen in Chart 7-1. Figure 7-2 shows the knuckle and ball joints added to the layout. If these ball joint locations are put on the layout, their coordinates would be as listed in Chart 7-2.

King Pin Inclination & Scrub Radius

If you draw a line through both ball joints and extend it to where it hits the ground plane, the *king pin inclination* is the angle that the ball joint line makes with the ground. Typically, this angle is

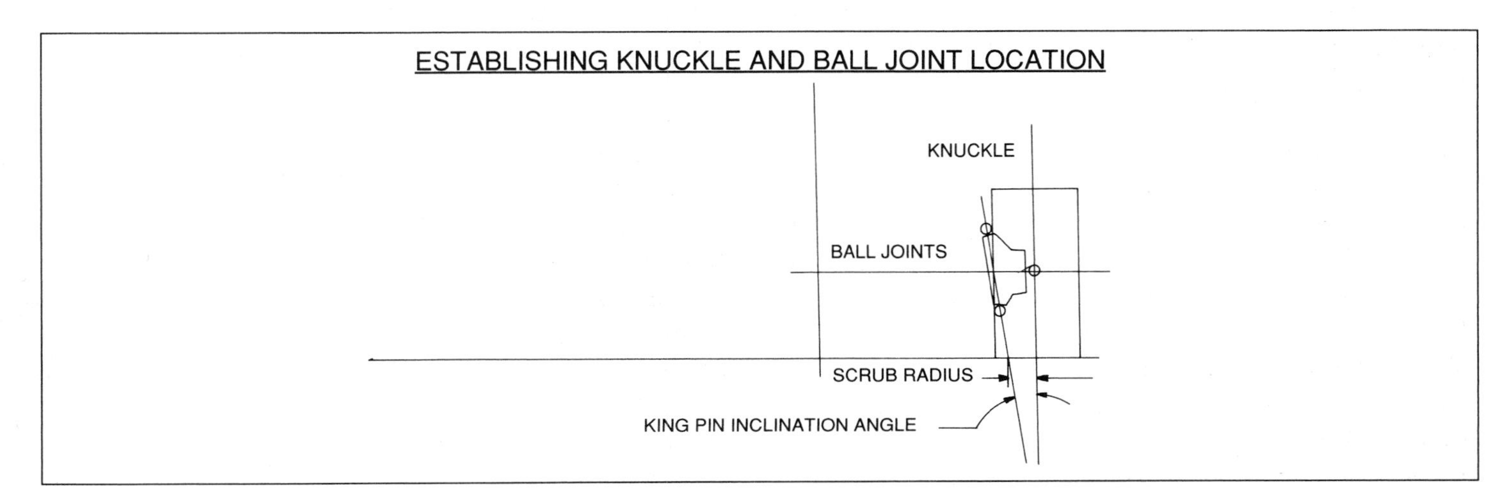

Figure 7-2. The knuckle is really the key component in a front suspension because it sets up where all the other components will be. When designing a front knuckle, it is best to keep the ball joints close to the wheel centerline to minimize the scrub radius.

Chart 7-2

BALL JOINT LOCATIONS

Location	X Direction Longitudinal Distance From Front Axle Centerline (inches)	Y Direction Height Above Ground (inches)	Z Direction Distance From Car Centerline (inches)
Lower Ball Joint	-.100	6.75	24.75
Upper Ball Joint	+.100	18.25	23.00
Steering Ball Joint	-4.00	7.25	25.00

NOTE: The dimensions in the "X" coordinate are negative if they are ahead of the axle centerline.

This example shows some positive caster at design and the steering arm ahead of the axle.

between 5 and 10 degrees. The *scrub radius* is the distance from the ball joint line to the centerline of the tire. There are significant handling and control advantages in reducing the scrub radius to the minimum. Any bump or cornering force that is applied to the tire can exert a twisting force on the steering that is proportional to the length of the scrub radius. If the scrub radius was zero, these twisting forces would be zero. Cars with zero scrub radius can usually be driven without power steering, because the twisting forces are gone and the tire easily rotates about the steering axis. The factors that increase scrub radius are: positive wheel offset, wheel and tire width, brake rotor width and the design of the knuckle. If any of these components can be made to minimize the scrub radius, there will be an improvement in handling, control and steering effort.

As can be seen on the layout, more king pin inclination will also reduce the scrub radius. Increasing the king pin inclination also helps to center the steering, so why not use a lot of king pin angle? The reason is another compromise situation. With excessive king pin angle, the tire tends to flop from side to side as it is steered. This causes the tire contact patch to run up the edge of the tire as it is turned. The above considerations and many experimental cars in the past, have shown that the best compromise in king pin angle is between 7 and 9 degrees (Figure 7-3).

Control Arm Length

Assuming you have established the tire/wheel, knuckle and ball joint locations on the layout, the length of the control arms is the next step in the design process.

Lower Control Arm—In general, it is best to make the lower control arm as long as possible. This reduces the angularity the ball joint must accommodate as well as slowing down angular change of the suspension members as they go through their travel. Long control arms cause clearance problems to the engine, exhaust, etc., so again you are faced with a compromise. Our example layout shows lower control arms 13.0 inches long, which is about the practical minimum for a suspension with 4.00 inches of jounce and rebound travel.

Figure 7-3. This photo shows a cast stainless steel knuckle tucked inside a wheel used on a NASCAR Winston Cup race car. With this compact design, it would be possible to have even less scrub radius if the wheel center was moved outboard.

Upper Control Arms—The length of the upper control arms can have a dramatic effect on the suspension system because it controls the movement of the roll-center during jounce and roll. In general, the upper control arm length will be between 50% and 80% of the length of the lower control arm. It is a popular belief that a shorter upper control arm is an advantage because it results in a faster camber gain at full jounce. Since this condition is only one aspect of the suspension movement, it is a mistake to compromise the total suspension design for this small advantage. With computer analysis, it is possible to make the upper control arm length exactly right so there is little change in the roll center location in either jounce or roll. Later in this chapter, I will describe the use of computers in optimizing suspension geometry design.

For our sample layout, I have chosen an upper control arm length of 7.00 inches. The combination of ball joint heights, tire diameter and lower control arm length need this upper control arm length to

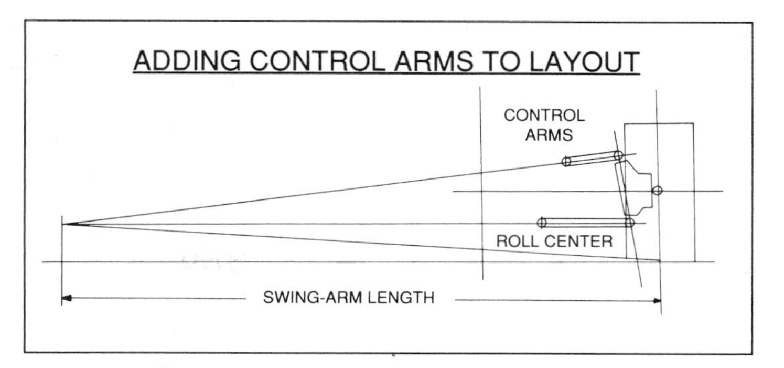

Figure 7-4. After the ball joint locations are established, the control arms can be added. Their location and length will determine the camber gain, instant center location, and the movement of the instant center as the suspension moves up and down.

provide a smooth wheel movement while controlling the roll center location. Figure 7-4 shows the layout with the control arms added.

Swing-Arm Length & Camber Gain

The *swing arm length* is the distance from the ball joint line to where the upper and lower control arms intersect. A longer swing-arm length results in smooth wheel movement but less camber gain. A shorter swing-arm length results in more camber gain but causes the wheel movement to become more erratic. Some enthusiasts might believe that a shorter swing-arm length is better, but by stroking the suspension through its travel, you can see the error in this assumption. If you positioned your suspension at full jounce you would see that the swing-arm length would become very short. This is because the upper arm is shorter, and its angle changes in jounce to move the intersection of the arms closer and closer to the center of the car. Numerous tests have shown that a swing-arm length at a design ride height of between 100 and 150 inches is a good compromise. Our example layout has a swing-arm length of 124 inches at normal ride height. At full jounce (4.0 inches) the swing-arm length reduces to just 30 inches. This means that at full jounce the front control arms are, in effect, giving us a swing-axle suspension (see page 39). Think what would happen if we started with a swing- arm length of 60 in. At full jounce, our effective swing-arm would be less than 30 in. and the suspension would be just like the rear of an old Volkswagen. The maximum amount of camber gain can be had with a swing-axle type suspension, but I don't know of any designers who want to use this arrangement on the front of

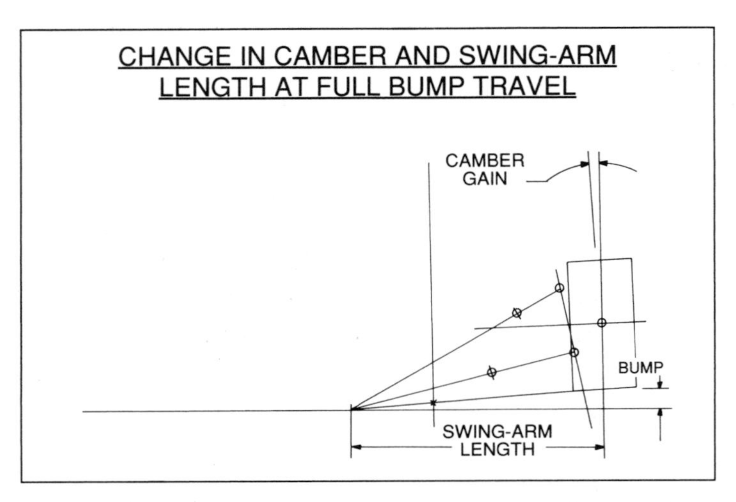

Figure 7-5. As the suspension compresses over a bump or in a banked turn, the camber increases and the swing arm gets shorter. Both of these factors help to maintain the proper camber angle of the outside wheel. Too short a swing-arm length can produce a jacking effect just as it does on a swing-axle front suspension.

their cars, so why arrange the control arms to give these same characteristics? As I explained earlier, suspension design is a compromise that requires evaluation of all the factors. Some camber gain is good, but that does not mean that more camber gain is better. With the 124-inch swing-arm length used in our sample layout, the camber gain is between .60 and 1.6 degrees per inch of jounce travel. At the full jounce position, there is a camber gain of over 4 degrees. Figure 7-5 shows how to determine swing-arm length and the camber gain.

Movement of Roll Center

As I explained previously, the roll center location can move both vertically and laterally if the upper control arm length is not correct (Figure 7-6). Since the roll center height has an effect on body roll, and on how much of the body roll is absorbed by the front suspension, it follows that if the roll center height changes when the car hits a bump, or when the ride height changes on a banked turn, the driver will notice a change in handling. If you keep the roll center height constant throughout the suspension travel, the handling will be more consistent. The same is true for roll center location across the car. Unless the upper control arm length is optimized, the roll center location will move laterally as the car experiences some roll angle. Again, if the roll center is moving across the car during its travel around a corner, the driver will feel a change in handling. If you can eliminate unpredictable changes in the way a car feels on the road, you or the driver will have more confidence in the car's handling and be able to run faster.

Front Suspension Pivot Points

Now that the desired camber gain and roll center heights have been determined, you can establish the exact coordinates for the control arms. These coordinates are needed to use a computer program to study the suspension characteristics during jounce and rebound travel. They are set in Chart 7-3.

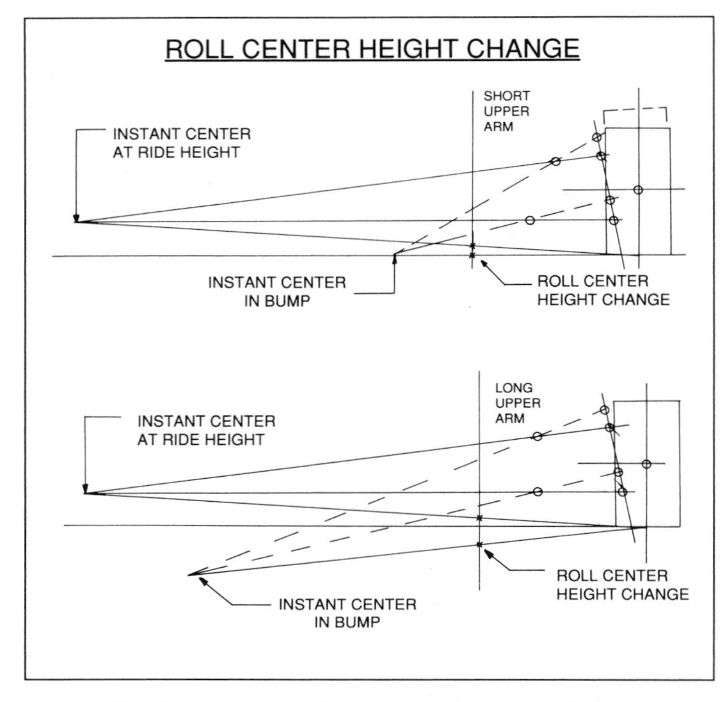

Figure 7-6. The roll center height can change as the suspension moves if the lengths of the control arms are not properly matched. A changing roll center height will make a car feel slightly unstable

COMPUTER ANALYSIS OF SUSPENSION GEOMETRY

The suspension pivot points established on the layout can be fed into a computer program to analyze how they interact as the suspension moves through its travel in jounce and rebound. It is interesting to note that there are only nine pivot points in the entire suspension and steering system. From these nine points, the computer program can print out 10 pages of data (Figure 7-7). We will only show the first page here, but here is a list of the information that is also available.

Roll Angle of Car
Camber Angle of Both Wheels
Roll Center Location Vertically & Laterally

Chart 7-3

Coordinates	X Direction Distance from Front Axle Centerline (inches)	Y Direction Height above Ground (inches)	Z Direction Distance from Car Centerline (inches)
Lower Control Arm			
Front Pivot	-1.500	5.900	11.750
Rear Pivot	+10.000	5.900	11.750
Ball Joint	-100	6.750	24.750
Upper Control Arm			
Front Pivot	+.100	17.120	16.150
Rear Pivot	+6.100	16.920	16.150
Ball Joint	+.100	18.250	23.000
Tie Rod Inner Pivot	-4.000	6.400	12.150
Tie Rod Outer Pivot	-4.00	7.250	25.000
Wheel Center	+.000	12.500	30.000

Each of these parameters is described with numbers for each 1/2 inch of wheel travel and each degree of roll. All the numbers can be plotted on paper to determine what the suspension geometry is doing under almost any imaginable situation.

The computer program used for this example is available from: Peerless Engineering, 840 Dahlia, Rochester Hills, MI 48307 (313) 651-5140.

Analyzing the Data

Here is how I would analyze the printed data that comes out of the computer shown in Figure 7-7. The first line of data shows the computer calculated dimensions that I selected on our sample layout. Shown are the true lengths of the control arms, the scrub radius, the king pin angle and length, the tie rod length and the track dimension. Although I had all this data on the layout, this shows that the computer agrees, so there is no mix up in transferring the data. The computer data sheet is arranged so you can see how the wheel position and geometry parameters change as the suspension is moved through its travel. Notice that there is data for wheel travel 4.00 inches up and 4.00 inches down. The positive numbers indicate the bump direction, or the wheel moving up in relation to the chassis.

Caster Angle—As can be seen from the column of numbers, our design geometry starts with just less than one degree of caster at ride height and increases to just over 1 1/2 degrees of caster at full compression. In rebound, it loses about 7/10ths of a degree of caster. The small amount of caster change is a result of having the upper control arm pivots angled upwards at the front. We angled these pivots to provide some anti-dive under braking, and the caster change is a by-product of this.

Although there is some change in caster, the total amount is so small that you would normally ignore it for most high-performance applications. However, slight changes in caster can cause a change in

SLA FRONT SUSPENSION GEOMETRY PROGRAM

FILENAME = HLAST
DATA TITLE = HLA SHORT TRACK11/
FOR THIS PROGRAM, X=AFT, Y=UP, Z=LEFT

THIS IS A FRONT SUSPENSION WITH THE WHEELS NOT DRIVEN

LOWER ARM COORDINATE			(X, Y, Z)	UPPER ARM COORDINATES (X, Y, Z)		
FRONT PIVOT =	-1.500	5.900	11.750	.100	17.120	16.150
REAR PIVOT =	10.000	5.900	11.750	6.100	16.920	16.150
BALL JOINT =	-.100	6.750	24.750	.100	18.250	23.000

TIE ROD INNER PIVOT = -4.000, 6.400, 12.150
TIE ROD OUTER PIVOT = -4.000, 7.250, 25.000
OTHER DESIGN PARAMETERS

JOUNCE TRAVEL = 4.000	CAMBER = .000	WHEELBASE = 101.000
REBOUND TRAVEL = -4.000	SLR = 12.500	C.G.HEIGHT = 18.000
% FRONT BRAKING = 60.000	WHL CNTR COORDINATES = 000	12.5000 30.150

LOWER ARM TRUE LENGTH	UPPER ARM TRUE LENGTH	SCRUB RADIUS	KINGPIN ANGLE	TIE ROD LENGTH	TRACK
13.028	6.942	4.373	8.653	11.634	60.300

RADIUS R3 = 26.83

WHEEL TRAVEL	CASTER ANGLE	CAMBER DEG.	TIRE SCRUB	TOE-IN DEG.	FRONT SWG. ARM	SIDE SWG. ARM
4.000	1.591	-4.343	-.264	-.154	-32.55	499.25
3.500	1.520	-3.491	-.211	-.130	-36.55	471.54
3.000	1.449	-2.741	-.163	-.108	-41.31	449.11
2.500	1.376	-2.083	-.119	-.089	-47.09	430.60
2.000	1.302	-1.512	-.081	-.070	-54.28	415.01
1.500	1.228	-1.021	-.050	-.053	-63.54	401.65
1.000	1.152	-.607	-.025	-.035	-76.09	389.95
.500	1.075	-.267	-.008	-.018	-94.32	379.49
.000	.996	.000	.000	.000	-123.82	369.89
-.500	.917	.194	.000	.018	-181.35	360.83
-1.000	.836	.314	-.010	.037	-350.11	352.01
-1.500	.754	.356	-.029	.055	-45629.80	343.09
-2.000	.670	.313	-.057	.073	322.87	333.73
-2.500	.585	.174	-.096	.090	150.02	323.46
-3.000	.497	-.076	-.143	.105	91.38	311.67
-3.500	.406	-.463	-.199	.115	61.02	297.41
-4.000	.312	-1.031	-.263	.116	41.83	279.09

Figure 7-7. Here's a sampling of a computer program available from Peerless Engineering that will enable you to see how the camber, toe, caster and swing arm length changes as the suspension moves through its travel. It's also possible to determine roll center height, anti-dive, understeer, roll angle, and many other design parameters with the program.

handling on speedway racers, so it might be better to forego the anti-dive feature for this application.

Camber Angle—The numbers shown indicate that our sample suspension design has a total camber gain of negative 4 degrees at the full bump of 4.00 inches. It should be noticed that the camber gain is progressive in that there is more gain between 3 and 4 inches of travel than there is between 0 and 1 inches of travel. Most enthusiasts think that more camber gain is better. But, the data shows that camber gain keeps increasing as the springs are compressed, so you can reach a point where the tire will be riding only on its inside edge. From a design standpoint, cars that run on flat tracks can use more camber gain. Cars that run on banked tracks should have less camber gain.

The amount of camber gain is dependent on the roll center height and, more importantly, on the length of the front-view swing-arm. A higher roll center height can actually decrease camber gain, so you can't just change the upper arm pivot location and expect your car to corner better. The difference in length between the lower and upper control arms determines how progressive the camber gain is. A shorter upper arm will make the camber gain more progressive, but too short an upper arm will cause radical changes in the roll center location during body roll and over bumps.

Tire Scrub—This column of numbers indicates how much the tire contact patch moves sideways as the suspension moves through its travel.Obviously, you want as little tire-scrub as possible.

Toe Change—The computer measures toe-in in degrees rather than in inches. Our sample suspension design shows a total toe-out of .154 degrees at 4 inches of bump. This converts to about .040 inches at full travel. The change in toe, as shown on the computer sheet, is called *bump steer*. Bump steer is how much and in which direction the toe changes during suspension travel, such as when going over a bump. If the outside front tire toes-out in bump, you have *bump understeer*. Generally, bump understeer feels better to the driver because the car only turns as much as he tells it to. The amount of the toe change, and whether it is toe-in or toe-out, is determined by the position of the inner tie rod pivot as compared to the tie rod pivots on the knuckle. What should be noticed here is how accurately the computer can determine changes in toe, without ever having to build and test the hardware.

Front-View Swing-Arm—The front-view swing-arm is the distance from the knuckle to where the upper and lower control arms would intersect if they were projected far enough. The length of the front-view swing-arm has a great influence on how much camber gain the suspension has. A shorter front-view swing-arm results in more camber gain. If you want the most possible camber gain, you could replace your double A-arm suspension with a swing axle suspension. This computer data shows that at ride height, the front-view swing-arm is 124 inches long. At full bump, its length reduces to 32 inches. Since this is about the length of a swing axle, all the problems of a swing-axle suspension will result if a shorter swing-arm length is chosen at ride height.

Side-View Swing Arm—The side-view swing-arm is the distance from the knuckle to where the upper and lower control arms would intersect if they were projected far enough in the side-view. Since our sample suspension design has very little anti-dive built in, the length of the side-view swing-arm is quite long. As more anti-dive is used, the side-view swing-arm length decreases, and as we discussed before, caster change will increase.

Roll Center Height—Although not shown in Figure 7-7, our sample suspension design shows that the roll center height is just below the road surface at ride height. As the suspension is compressed, the roll center height decreases at about the same rate. This means that the height of the roll center does not change much in relation to the rest of the chassis. As discussed in Chapter 3, the roll angle of the car is dependent on the distance from the roll center to the center of gravity. If this relationship stays constant as the suspension is compressed (like during the banking on a high speed oval), the driver will not notice a change in handling as he runs from the flat straightaway to the banked turns.

Reducing the distance between the roll center height and the center of gravity height will reduce the roll angle, which leads many racers to want to raise the roll center. But a higher roll center causes jacking effects and erratic suspension movements. In some cases, raising the roll center height can also reduce camber gain. Most successful cars have the roll center height between 1.00 inches below ground to 3.00 inches above ground, so I would recommend you stay in this range to keep all the compromises at a reasonable level.

Percent Anti-Dive—The computer data showed that the anti-dive values of our sample suspension range from 5% to 7%. As discussed previously, this is a slight amount, because we didn't want to introduce any more caster change. Some production cars use as much as 30% of anti-dive because they want to keep softly sprung cars level during heavy braking. Most race cars are light enough and low enough that they don't have a real need for anti-dive effects. Figure 7-8 shows a layout of the anti-dive geometry.

Percent Understeer—Percent understeer is a calculated factor that indicates how much the suspension design and steering geometry will make the car want to understeer in the turns. Our sample suspension design on the program (not shown) indicated about 2% understeer, which is a small amount. From a practical standpoint, this value is not important by itself because it doesn't consider all the aspects of how the car is set up for high performance driving.

BUMP STEER

Included in the computer analysis was data relating to the steering geometry of the design. The data showed that the total change in toe-out for the full 4.00 inches of bump travel was only .040 inches. In the full rebound travel of 4.00 inches, the toe-out was less than .030 inches. From a design standpoint, this would be considered a pretty good layout. If the parts were all made to the exact dimensions, you could check the bump steer and it would closely agree with the computer data for change in toe.

Bump steer is how much and in which direction the toe changes during suspension travel. If a chassis is designed and set up correctly, there should be very little bump steer. Even with a well-designed chassis, small changes to steering component location can cause significant changes in bump steer. In order to be sure your car is set up correctly, you might want to measure the bump steer. See Figure 7-9 for how bump steer is determined on a layout.

Measuring Bump Steer

Since bump steer is the change in toe through suspension travel, you need a way to measure toe change at different wheel travel positions. You can buy a special fixture that makes this a simple operation, or you can improvise with a simple method that delivers slightly less accurate results. Make sure the wheels are pointing straight ahead and that the steering gear is in mid-position. This is important, because very small changes in the steering pivot points can have significant effects on the bump steer characteristics. Place a piece of plywood on edge against the front tire. With the suspension at ride height, the board should touch both the front and the rear edge of the tire. Remove the springs and use a jack to move the suspension into the bump position. When the bottom of the board is kept stationary on the ground, you can tell any change in toe by looking at how well the top of the board aligns with the front and rear edge of the tire.

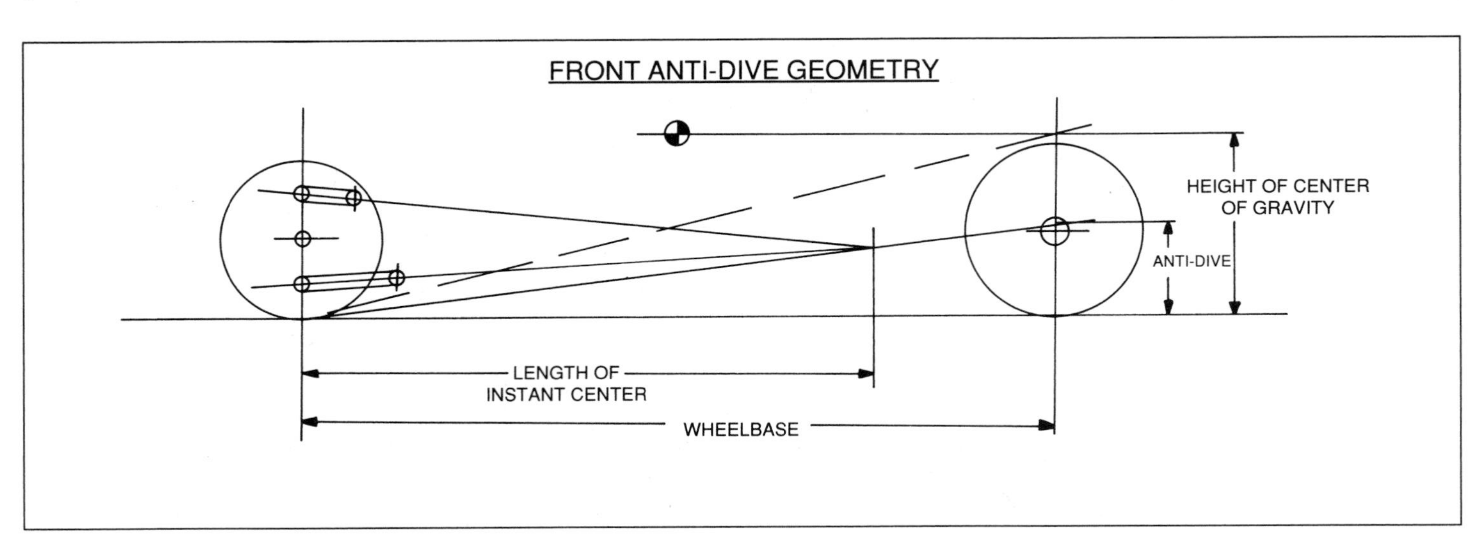

Figure 7-8. The arrangement of the control arms in the side view determines the anti-dive characteristics of the front suspension. The computer data in Figure 7-7 of our sample suspension shows anti-dive between 5 and 7 percent. We wouldn't want to go any higher because we don't want to have any more of a change in caster for this application.

The maximum anti-dive on street cars is usually less than 30 percent, to keep them level during heavy braking. But many race cars use zero anti-dive, because they have such a low center of gravity. Photo by Michael Lutfy.

If there is a .030-inch air gap at the rear of the tire, your car has this amount of bump steer, and it's in the understeer direction.

The same basic process is used with professional bump steer equipment, but of course the results are much more accurate. See Figure 7-10 for how to measure bump steer on a car.

Bump Steer Amount & Direction

Exactly how much bump steer you need on your car is like most suspension settings—a compromise. It is common to set the bump steer so that the front wheels toe-out on a bump. This will make the car feel more stable, because the car will not turn any more than the driver asks. To understand this effect, picture what would happen if your car had toe-in on bump. As the driver would start a turn, he would feed in a certain amount of steering angle. As

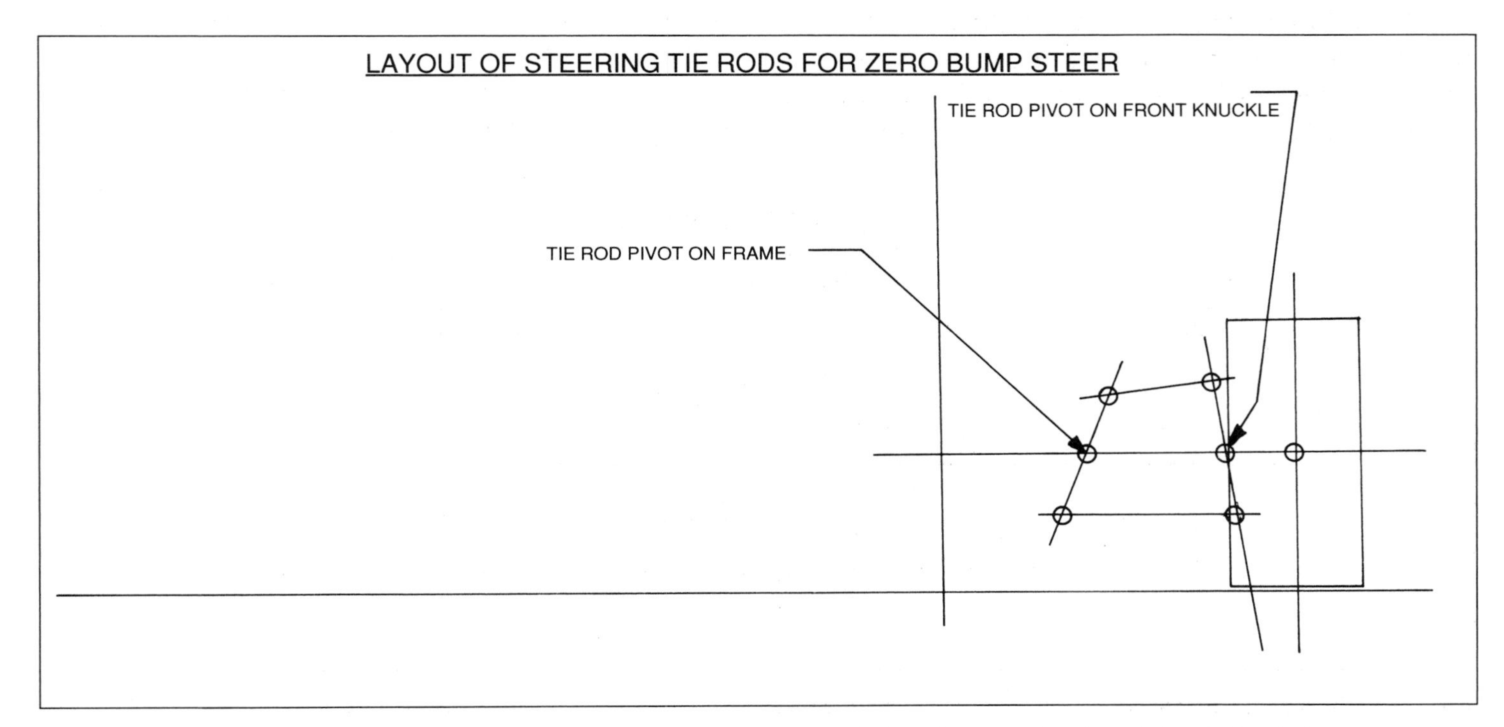

Figure 7-9. The amount and direction of the bump steer is controlled by the location of the steering tie rods in relation to the control arms. With a computer, we can design for zero bump steer or any amount in either direction.

the car built up *g*-forces, the chassis would roll and the outside suspension would compress in the bump direction. If the car had toe-in on bump, the front wheels would start to turn more than the driver asked and his turn radius would get tighter. This would require the driver to make a correction and upset the car's smooth approach into the turn. The outside tire is considered in this analysis because it carries most of the weight in a turn.

Assuming that your car has the bump steer set so that there is toe-out in the bump direction, the next consideration is *how much* toe-out. If the car has too much toe-out in bump, the steering can become imprecise, because the suspension will tend to negate what the driver is doing with the steering wheel. Also, if there is too much bump steer, the car will dart around going down the straightaway. A reasonable amount of bump steer would be in the range of .010 to .020 per inch of suspension travel.

How to Adjust Bump Steer

On a typical double A-arm suspension, the knuckle moves in radii set up by the length of the control arms. Since the upper arm is usually shorter than the lower arm, the top of the knuckle moves in a tighter radius than the bottom of the knuckle. By placing the steering inner pivot at exactly the correct position, it is possible to make the wheels go through their travel with no change in steering angle. This would be zero bump steer and there would be no toe change. By moving the steering pivots, it is possible to change the amount of bump steer and the direction of the bump steer from toe-in to toe-out.

Rate of Change—It is also possible to adjust the rate of change of bump steer. If the inner steering pivot is too far inboard (a longer tie rod), the rate of change of toe-out will be reduced. For each suspension layout, there is an ideal length tie rod that will give the correct rate of toe change. This length is usually established by the chassis builder, so you shouldn't have to consider its effects unless you are the chassis builder, or if you have difficulty getting your bump steer adjusted. On a car that has the steering gear ahead of the front suspension with the correct length tie rod by design, more toe-out will result in bump if the inner steering pivot is raised or if the outer steering pivot is lowered. These adjustments are the ones normally used to adjust bump steer. Each chassis has its own adjustment characteristics, but the concept will be the same.

Bump steer has a big influence on the driver's ability to control the car, so it is an important part of the design and setup of your suspension.■

Figure 7-10. Regardless of how well the steering gear and linkage gear are designed, it is a good idea to check the actual bump steer on the finished car. Bump steer can be measured by observing how much the toe changes as the suspension is moved up and down. Place a flat board against the tire (in this case we used a creeper), then take a jack and lift the wheel assembly up and down. By looking closely at the two photos, you can see the amount of toe change.

BUILDING A FRONT SUSPENSION

We have discussed most of the design aspects of a front suspension, but engineering evaluation and analysis doesn't stop there. In assembling the components for a front suspension, you have to evaluate each piece to determine if it is strong enough to suit your needs while at the same time providing the lightest possible weight. This is obviously a compromise that is complicated by the fact that you really don't know if any given piece will be strong enough until it is tested. This is where experience becomes so valuable. If parts are known to survive the rigors of endurance racing, such as a 500-mile event, then you can use them with confidence in your car.

Tires

The tires have more effect on how well a car handles than any other component associated with the front suspension. Because they have such a big effect, their performance characteristics are covered in Chapter 1. We use Goodyear tires for our test and development because they are "state-of-the-art." Goodyear's ongoing racing research and development work assures that their tires are always at the leading edge of the available technology. Which compound and construction to use is best decided after testing.

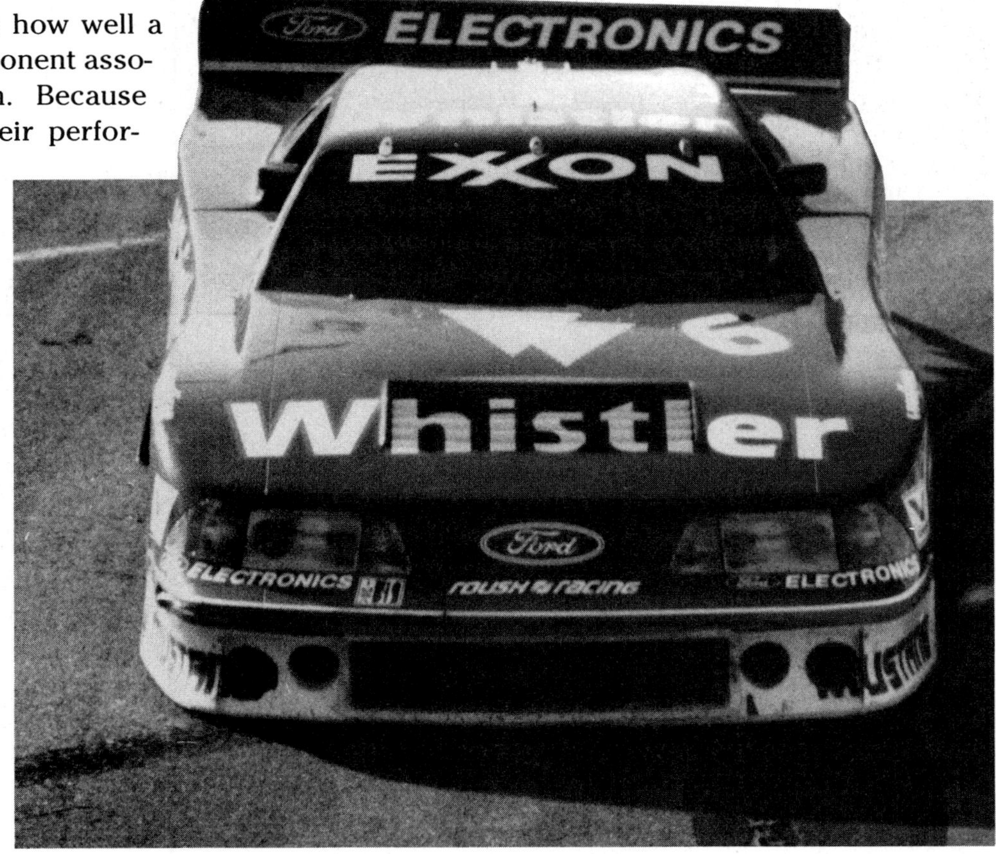

You can design the perfect suspension for your car on paper, but even the best design won't be enough to overcome the faults of weak materials. Suspension components, especially those used for racing purposes, must be strong yet lightweight. If you use components known to withstand the rigors of racing, especially endurance events, then you can use them with confidence on your car. Photo by Michael Lutfy.

Wheels

In order to reduce your car's front scrub radius, you should use wheels with zero or negative offset. Zero offset means that the mounting surface of the wheel is on the same centerline as the tire. Since a 10-inch wide wheel is really 11.0 inches across the flanges, the *backspacing* for zero offset wheels is 5.50 inches. Backspacing is the distance from the wheel mounting surface to the innermost flange edge.

If you had a wheel with 1.00 inch of negative offset (mounting surface outboard of the wheel centerline)

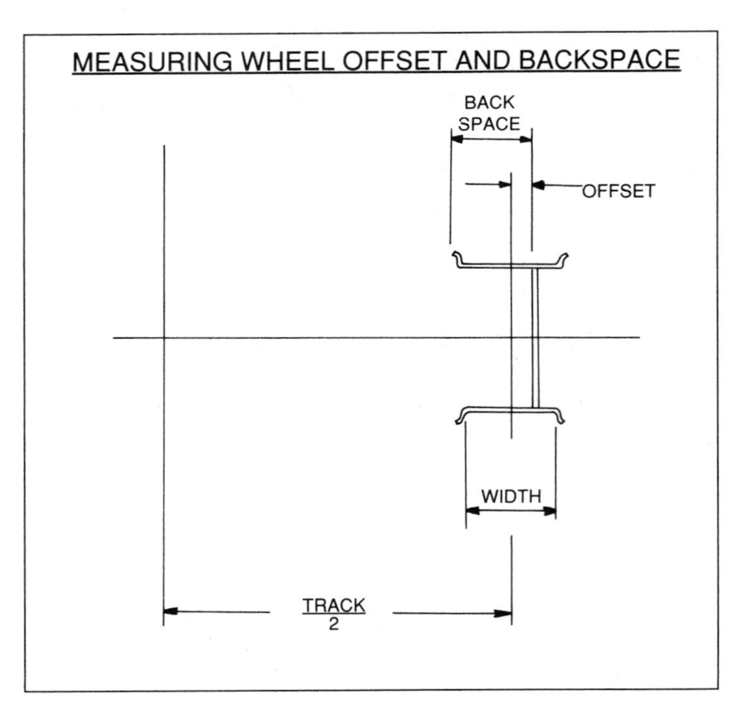

Figure 8-1. This diagram shows the relation of the wheel dimensions commonly used in the aftermarket industry. Note that the overall width of a typical wheel is 1.00 in. larger than the tread width because of the flanges.

the backspacing would be 6.50 inches. Figure 8-1 shows where wheel offset and backspacing are measured from.

Three-piece component wheels allow for building a wheel with varying amounts of offset, so these are preferred. Three-piece wheels are also usually the lightest and the most expensive. If your budget dictates steel wheels or cast aluminum wheels, you can still measure different types to find the wheel offset you need.

Using three-piece wheels allows for the flexibility to build wheels with a variety of diameters, widths and offsets. The current selection of wheels available are lightweight, attractive and expensive.

This knuckle is made from cast stainless steel, which results in a lightweight part that is quite rigid and compact. The hub is made from 4130 billet steel and heat-treated for improved strength.

Knuckle & Hub Assembly

In order to save front-end weight, I recommend the use of a newly designed knuckle and hub made from cast stainless steel. This material provides excellent strength, and because it can be cast in thin sections, it is possible to design it for rigidity and lightweight. A complete knuckle-and-hub assembly with bearings and seals weighs about 18 lbs. This knuckle design allows you to place the hub further inside the wheel, reducing the scrub radius and improving steering precision.

Mounting brackets are incorporated in the cast knuckle for the brake calipers and for the cooling hose fitting. Dual steering arms are also incorporated, so it can be used on either front-steer or rear-steer front suspension. This provision also allows it to be used on either the right side or the left side. The knuckle and hub shown in our example nearby is available from: VSE, 23865 Fairfield, Carmel, CA 93923, (408) 649-8423.

Wheel & Brake Pilots

Because the front hubs rotate inside the front knuckle, a one-piece wheel-pilot, wheel-spacer and brake rotor mounting hat can be used. This part simplifies the design and saves a little weight. Various wheel offsets and track dimensions are possible with changes in how deep the machining for the hub pocket is made. Any diameter rotor pilot can also be machined on the same basic casting.

This one-piece aluminum pilot serves three functions; as a wheel pilot, brake rotor pilot, and wheel spacer. The thickness of the spacer can be machined to provide custom wheel locations.

Control Arms

The construction of the upper and lower control arms should be done with consideration for making them as rigid as possible. The ball joints should be proven components and their attachment to the arms should be secure. A previous chapter explained the importance of controlling bushing deflections and their effects on handling.

Mounting—When mounting the control arms to the chassis, it is important that the brackets be made rigid. Any flexing of the control arm brackets will result in deflection and/or vibrations when the control arms are loaded during hard cornering. It is also important that the whole front of the frame doesn't deflect under loading, so care must be taken to fully triangulate this structure. Chapter 12 explains how to build a rigid frame structure.

Fabricating control arms from steel tubing is common practice. It results in strong and rigid parts that can be built to almost any design requirement.

Coil-over springs and shocks are popular because they are simple and lightweight. They also provide some adjustment which makes turning easier.

Springs/Shocks/Jounce Bumpers

From the sample design layout in Chapter 7, you know that you need 5.00 inches of shock travel to give 8.0 inches of wheel travel. This results in shocks with an extended length of 16.0 inches. These units are available in various dampening levels, and can be had with rod-end attachments at both ends. Shocks should feature a gas-cell design to prevent fade under racing conditions. The shocks can also be supplied with coil-over springs in various rates, so you can test to determine the optimum setup for your car.

Good shocks should also include a jounce bumper, which is really a variable-rate spring that lets you run a soft suspension spring rate for maximum road-holding. With the use of a jounce bumper, the suspension won't bottom out under severe bumps, because its variable-rate feature comes into play as the suspension is compressed. A jounce bumper can increase the ability of a suspension to absorb a severe bump while retaining a soft spring rate under normal conditions.

Note that a much higher spring rate would be required to prevent bottoming out if a jounce bumper was not used. Also note that the jounce bumper does not come into action until the suspension has already compressed.

It is very important to provide a rigid mounting for any steering gear. Any movement of the steering gear will result in imperfect feedback to the driver, and the car will feel difficult to control.

Steering System

The steering system on any car has a big effect on how comfortable a driver will feel while driving "at the limit." It is important to have a very precise steering system. There are a number of areas to consider when selecting a good steering system.

Front vs. Rear Steer—Front steer means the steering arms are ahead of the axle centerline, and rear steer means they are behind it. If all the components in the steering system, including the frame mounting, were made rigid and free of play, it would not matter whether a car had front or rear steer. The driver input and response would be the same.

In the real world, however, components are not always free of play, and they bend or deflect when they are subject to load. These deflections can have a significant effect on how the car reacts to the driver's input. Understanding how these deflections can affect the car's steering response is the secret to making your car feel comfortable during cornering. In general, a front-steer car will be more comfortable to drive, because any deflection of the control arms (which are higher loaded than the steering gear) will result in a small amount of understeer. This handling condition is easier for a driver to adjust to, so he will feel more comfortable with front steer.

Because of the very large forces that can be generated by a worm-and-sector steering gear, it is very important to provide rigid frame members to support the gear. If the frame bends when the steering is turned, the driver will not get an accurate feel for how the car is handling. This photo shows proper mounting.

Rack & Pinion vs. Worm & Sector—If both steering systems are rigid enough to handle the loads applied to them, there should be no difference in driver feel between a rack-and-pinion steering box and a worm-and-sector box. Each is just a gearbox, so if the lash is properly adjusted, one won't be any better than the other. Rack-and-pinion steering can save about 50 lbs. over a worm-and-sector system, so this is why it is used more often.

Steering Gear Mountings—Regardless of whether you have front steer or rear steer; manual or power; rack-and-pinion or worm-and-sector; the rigidity of your steering gear mounts is extremely important. If your rack-and-pinion uses rubber mounting insulators, replace them with solid ones. If your frame flexes where the gear is mounted, reinforce it. Whenever the steering gear can move under loading, the car will react differently from what the driver expects, and he will not be comfortable driving it at the limit.

You can test the rigidity of your car's steering gear mounting by turning the steering from lock-to-lock while the car is parked on the concrete. If the frame bends and the steering gear moves under loading, you need more rigid mountings.

Like most parts of a high performance car, the effectiveness of a stabilizer bar system is dependent on the rigidity of the mountings and linkages that connect it to the suspension.

Stabilizer Bar & Linkage

In order to take advantage of soft spring rates and a low roll center, your front suspension will need a very effective front stabilizer bar to limit the roll angle. Limiting the roll angle will also reduce the need for excessive camber gain, so you will be able to use a long swing arm length. A long swing arm length will give smooth and consistent transitions. There are four factors that are important in building an effective front stabilizer bar.

Bar Diameter—The torsional stiffness of a stabilizer bar increases with the diameter to the 4th power (see page 15). This means that a 1.00-in. dia. bar will be 16 times as stiff as a 1/2-in. dia. bar. You can also use a 1 1/2-in. dia. bar made from 1/4-inch wall tubing, which is lightweight yet offers good stiffness. It is also easy to fabricate.

Lever Arm Length—The torsional stiffness of a stabilizer bar increases as the lever arm length is shortened. This means that a bar with a 6.00-inch lever arm will be twice as stiff as a bar with a 12.00-inch lever arm. The practical limit on shortening the lever arm length is about 6.00 inches. Any shorter, and the angularity of the linkage system will not allow full suspension travel.

Links & Mounts—The stiffest stabilizer bar will not be effective if its links and mounts are not rigid and tight. Any movement of the bar that is not immediately transferred to the suspension will be lost and the bar will not be effective. Use tie-rod ends for your linkage, and mount the bar to the frame with aluminum blocks and nylon sleeves.

Mounting Location on Control Arm—Just as the wheel rate-to-spring rate ratio is determined by where the spring acts on the control arm, the effectiveness of the stabilizer bar is also dependent on where the link attaches to the arm. The front stabilizer bar's effectiveness is increased as the link mounting point is moved closer to the lower ball joint. Mount yours as close as is practical to make your total system as effective as possible. ■

LIVE AXLE REAR SUSPENSION DESIGN

9

A live axle rear suspension is one where both rear wheels are mounted on a rigid axle. Because the whole axle moves as a unit, and because it moves whenever either wheel hits a bump, it is called a *live axle*. Live rear axles are used on front-wheel-drive cars and on rear-wheel-drive cars. Obviously, a differential is needed on a rear-wheel-drive car (Figure 9-1).

Advantages

The advantages of a live axle are simplicity and rigidity. This translates into less cost and an easier installation. The other advantage of live axle rear suspensions is that they have been around longer, so there is more information available on how to make them work correctly. A well-designed and well-developed live axle will beat a poorly designed independent rear suspension, even on rough roads. On smooth roads, it is usually difficult to see any advantage for an independent rear suspension. In the 1980's, the SCCA allowed its Trans-Am cars to run either a live axle or an independent rear suspension. The cars raced on relatively smooth tracks and there was no apparent performance advantage for cars with independent rear suspension.

Disadvantages

The major disadvantage of a live axle is its inability to allow each rear wheel to follow the contours of a rough road. Most of this inability comes from the unsprung weight of the differential. Some cars use a de Dion rear suspension, which is a live rear axle with U-jointed drive axles, so the differential can be mounted on the chassis to solve this problem (Figure 9-1A). This system is even more complex than a regular independent rear suspension, so it is not widely used. While a live axle rear suspension might not be the best textbook solution, it can be

In the 1980s, SCCA rules in the Trans-Am series allowed for the use of either a live axle or independent rear suspension. At smooth tracks, such as the now-defunct Riverside Raceway in California, cars equipped with IRS did not appear to have any performance advantage. Photo by Michael Lutfy.

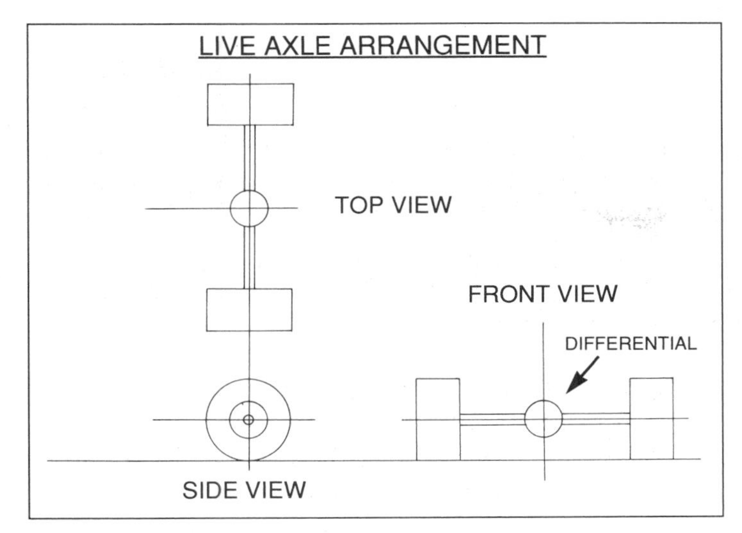

Figure 9-1. A live axle is really a beam axle at the rear that includes the differential. It is called "live" because the whole unit bounces in its entirety in relation to the frame. Although it is a single device, there are a variety of ways to connect it to the chassis.

made to work very well on smooth roads. And, because it has been so well developed, you can install one without too much trouble.

DESIGN REQUIREMENTS

Before discussing the various types of live rear axle suspensions and their advantages and disadvantages, let's first evaluate the design criteria, which will help make selection easier. There are a variety of ways to design a rear suspension using a live axle.

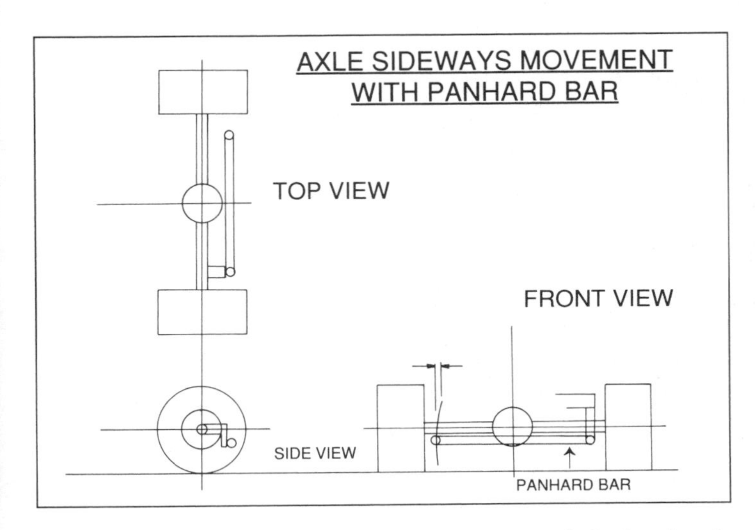

Figure 9-2. One of the ways to locate a live rear axle in the chassis is with a Panhard bar. One end of the bar attaches to the axle housing and the other end is connected to the frame. This is a simple and effective solution and the slight sideways motion is not a problem if the bar is kept as long as possible.

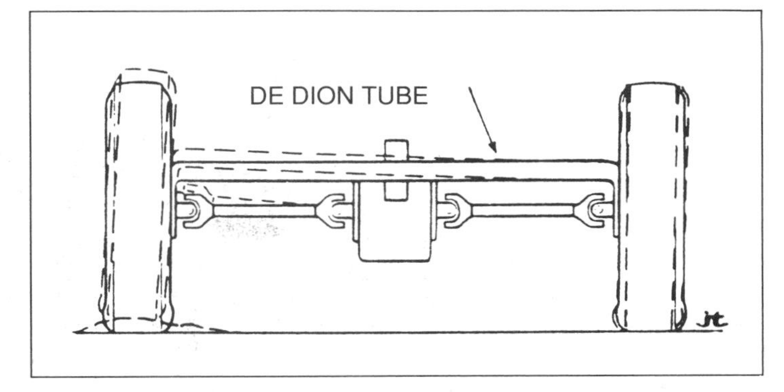

Figure 9-1A. A de Dion rear suspension is basically a live rear axle with U-jointed drive axles, so the differential can be mounted on the chassis. The complexity of this system, however, makes it impractical for high performance use.

Lateral Control & Roll Center Height

Lateral control is simply how the rear axle is kept in side-to-side alignment with the chassis as the rear suspension moves through its travel. There are a number of ways to provide this control, and the following are some of the more popular ones.

Panhard Bar—A Panhard bar is simply a link between the axle and the frame which controls the side-to-side location of the rear axle. The advantages of a Panhard bar are that it is simple, effective and lightweight. The disadvantages are that it must be as long as possible to minimize the slight side-to-side variations that result from the arc scribed by the bar (Figure 9-2). This slight variation has no real adverse effect on the car's cornering capability, but the driver does feel it so it can affect his performance. When a long Panhard bar is used, it must go

When using a long Panhard bar with extended bracketry, you must make sure the brackets are strong, rigid and yet lightweight. This means using high quality materials and involves excellent fabrication skills. Also, the bar should be mounted as low on the axle as possible, to achieve a low rear roll center.

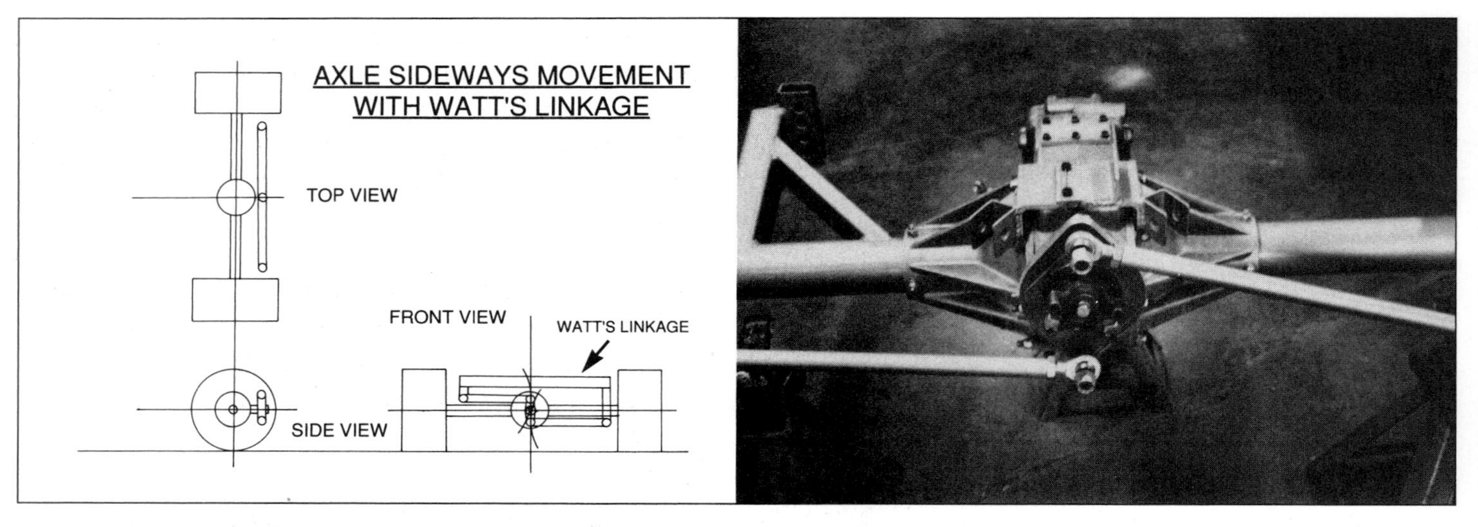

Figure 9-3. A Watt's link provides straightline, vertical movement between the axle housing and the frame. This precision means a more complex system that is probably not much of an improvement over a simple Panhard bar.

around the differential, and this requires extended bracketry. Making these brackets strong, stiff and lightweight is an engineering problem.

The roll center of a rear suspension that uses a Panhard bar is at the height where the bar attaches to the axle. A lower roll center at the rear makes a car handle more consistently, so I recommend mounting the bar as low as possible.

Watt's Linkage—A Watt's link eliminates the slight side-to-side variations that happen with a Panhard bar (Figure 9-3). Because a Watt's link gives straight line control of the axle, it is a better system from a design standpoint.

However, it is a more complex system, and because the roll center height is at the main pivot point, it is more difficult to arrange the mechanism at the best height. It is also more difficult to change the rear roll center height because the whole mechanism must be moved. If the Watt's link is mounted behind the axle, the bracketry must be very strong and stiff to absorb the loads without bending. Remember the loads on any rear axle lateral control system can be over 2000 lbs., with peak loads as high as 10,000 lbs.

Lower A-Arm—One way to provide lateral control of the rear axle is to replace the normal lower control arm with a larger lower A-arm (Figure 9-4). This design gives the axle straight-line location, while providing some of the linkage needed to locate the axle fore and aft. It provides a low roll center, although it does not allow for changing the height of the roll center easily. This design requires a very large rod-end joint at the axle, because it alone must carry the loads, especially when in bending. This arrangement is good from the design standpoint if you can make all the parts strong enough.

Lower Angle Arms—The same geometry and control characteristics of a lower A-arm can be achieved by using two lower arms angled to meet at the axle centerline (Figure 9-5 & 9-6).

The advantage of this system is that the rod ends are loaded in shear rather than in bending. This system also provides some opportunity to adjust the height of the rear roll center.

Bump & Roll Steer

Bump steer and roll steer are really the same thing. What they refer to is the amount and direction that the rear axle might cause the car to steer as

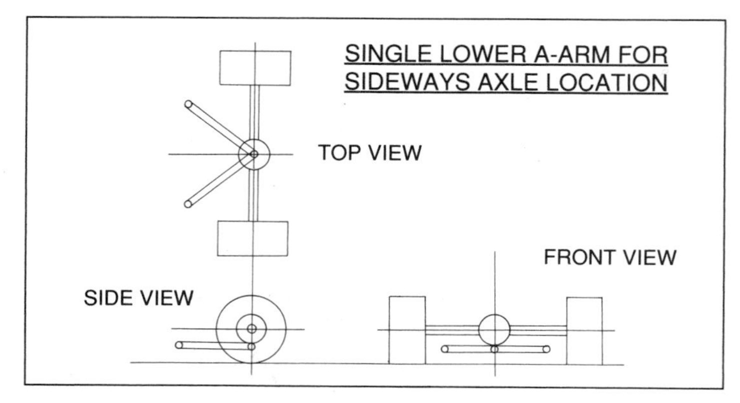

Figure 9-4. The rear axle housing can be located sideways to the frame by using a single large A-arm as one of the fore-aft links. This is a simple and effective system, but it requires a very large joint at the axle housing to absorb the cornering forces.

Figure 9-5. This photo shows the use of angled lower arms to locate an axle housing sideways in the car. This system is simple and lightweight. Various versions of this basic system are used on many production cars.

it moves through its travel. Even slight changes in the alignment of the rear axle will cause big changes in the direction the car will go. The preferred design characteristics are very little roll steer, and if there is any, it should be in the understeer direction. To accomplish this, it is necessary to arrange the rear suspension links so that the rear axle points to the left as the body rolls to the right and vice-versa. On a left turn, the body of the car will roll to the right. With the correct rear suspension geometry, this should cause the axle to point to the left, which will make the car turn right. This is called *roll understeer*, because it makes the car turn less as the body rolls. With *roll oversteer*, the car would turn more as the body rolls, and the two factors would augment each other to the point where the driver would have to steer away from the turn to keep on his intended path. In general, a rear suspension will have roll understeer if the roll axis tilts down at the front (Figure 9-7).

Rear Axle Steering & Alignment—The alignment and steering characteristics of a car's rear axle can have a dramatic effect on how the car handles and on how well it gets the power to the ground. If you have ever driven a forklift truck or if you have tried to race in reverse gear, you know first-hand how sensitive a vehicle is to changes in the steering angles and alignment of its rear wheels. Just because your car has a heavy-duty solid axle doesn't mean that the rear wheels don't steer your car. It is normal for rear-wheel toe-in and camber to change between

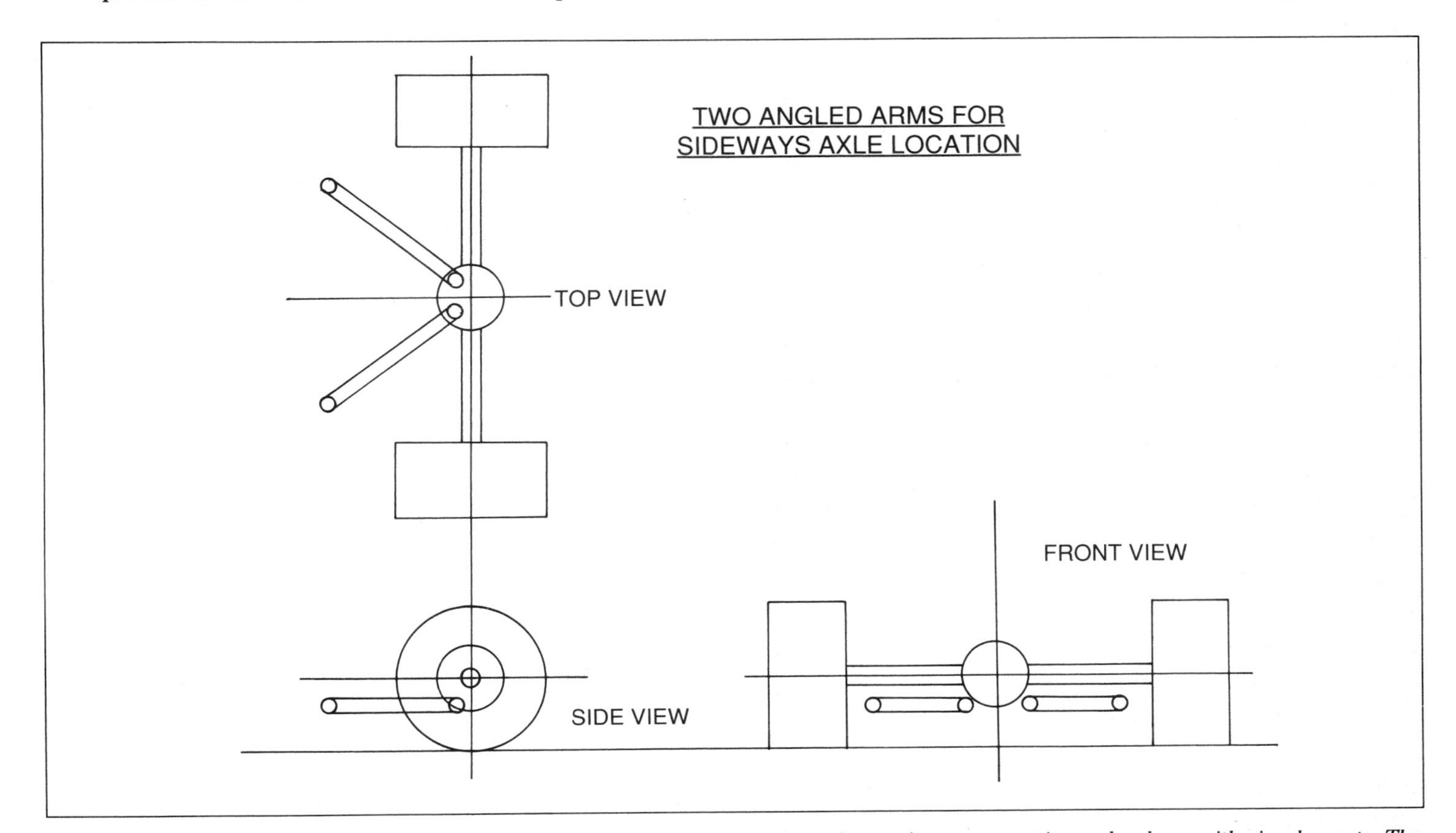

Figure 9-6. The use of angled lower arms serves the same function as a large lower A-arm except it can be done with simple parts. The advantage of two angled lower arms instead of a large A-arm is it eliminates the need for a large high-strength joint at the axle housing.

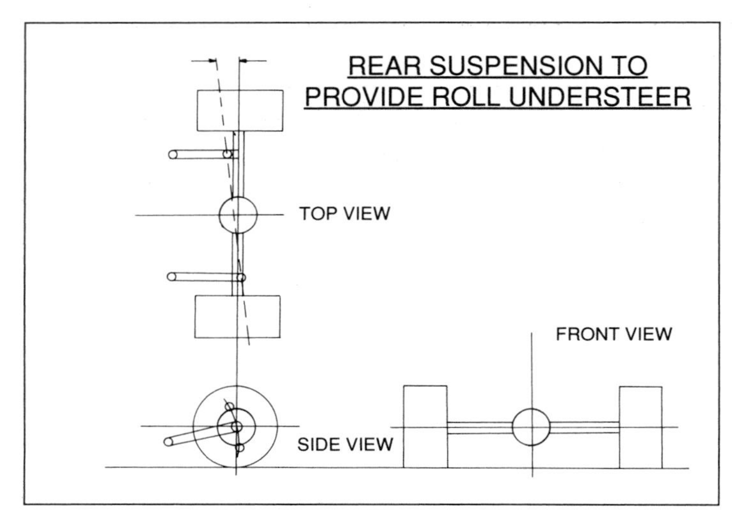

Figure 9-7. By angling the effective control arm system in the side view, it is possible to steer the rear axle as the body and chassis rolls in relation to the axle housing. It is usually best to make the rear axle steer to the left on a left-hand turn, because this makes the car turn less for roll-understeer.

one and two degrees as a car moves down the road. The basic problem is that the solid rear axle is not really rigid. The rear axle housing can flex just like any other member. If you recognize that the flex problem exists, you can take its effect into consideration when designing and setting up your car.

The best rear suspension design is only effective if all the parts are strong and rigid enough. Any deflections, or looseness of the parts, will have a greater effect on the car's handling than will its design geometry.

Anti-Squat

When a car accelerates forward, there is weight transfer from the front of the car to the rear. This weight transfer is dependent on the weight of the car, the height of the center of gravity and on the length of the wheelbase. Because of the springs, the weight transfer can often be seen at the rear of the car as it "squats" during hard acceleration.

It is possible to arrange the rear suspension links so that the driving force of the rear axle counteracts this squatting force. This characteristic is called *anti-squat*. Anti-squat can counteract the squat force to keep the rear of the car level, and it can be made strong enough to actually raise the rear of the car during acceleration. An example would be most drag race cars, which have enough anti-squat characteristics built in to actually raise the rear of the car under acceleration. Because any force that can raise the rear of the car will need to have an equal and opposite force pushing against the pavement, you can use anti-squat to increase the tire loading during acceleration. If anti-squat can increase the tire loading without adding to the weight that must be pulled around the corner, you will have more available cornering power at the rear of the car. With more cornering power at the rear, the driver can get on the throttle sooner without oversteer. This will result in faster cornering, so having more anti-squat characteristic is desirable.

Calculating Anti-Squat—You can determine the amount of anti-squat your suspension has by making a scale drawing like the one shown in Figure 9-8. You can determine the fore-aft location of your car's center of gravity by measuring the front and rear weight. It is more difficult to find the height of the center of gravity, so for most analysis, it is acceptable to assume a known center of gravity height based on other cars. Production cars usually have the center of gravity between 20 and 22 inches above ground. A typical oval short-track race car will have its center of gravity between 18 and 20 inches above ground. The wheelbase dimension and the locations of all the rear suspension pivot points can be determined by measuring your car. Be sure to put them on your drawing to scale so the relationships remain true. After the points are established, connect the control arm pivots and extend the lines forward. Where the upper and the lower arm lines intersect is the *instant center* of the rear suspension in the side view. The instant center is that point around which the linkage can be assumed to react. This point changes as the suspension moves up and down, but at any given instant, this is its *effective center*. If this point is on a line connecting the rear tire contact point with a point where the center of gravity height meets the front axle centerline, your car has 100% anti-squat; if the intersect point is below this line, you have less than 100% anti-squat; and if the intersect point is above this line, you have more than 100% anti-squat. The amount of anti-squat is proportional to the height above and below the 100% line.

Chart 9-1 is an example of how anti-squat would change rear tire loadings, cornering power, and traction available for acceleration. You can see that if the rear tire loading can be increased with anti-squat effects, there will be more rear tire traction available

Chart 9-1

	Without Anti-Squat		With Anti-Squat	
	Left Rear	Right Rear	Left Rear	Right Rear
Static Weight on Tires 3000 lb car 50% weight on rear 50% left side weight	750	750	750	750
Weights Due to Cornering Load Weight Transfer C.G. Height : 20 in Track Width: 60 in Cornering at 1.0 *g*s	250	1250	250	1250
Weights Due to Acceleration Induced Weight Transfer C.G. Height: 20 in. Wheelbase: 100 in Acceleration of .50 gs	400	1400	400	1400
Weights Due to Anti-Squat Effect Assume 200 lbs.	400	1400	500	1500
Tire Traction Available From Sample Tire Curve page 2	600	1200	700	1250
Total For Rear of Car	1800		1950	
Tire Traction Needed to Corner at 1.0 gs	1500 lbs		1500 lbs	
Tire Traction Available for Acceleration	300		450	

Most drag racing cars have enough anti-squat characteristics built into their car that the rear actually raises under acceleration. Photo by Michael Lutfy.

for acceleration coming out of the corners. This means that the car will be able to use more throttle without inducing power oversteer. The advantage of using anti-squat to tune corner exit handling is that it does not require losing any corner entry speed to achieve it. All racers know that the sooner the driver can put the power to the ground coming out of a corner, the faster the car will be down the straightaway. Using anti-squat allows this to happen sooner. The type of rear suspension and its adjustment determines how much anti-squat a car has.

Swing-Arm Length & Brake Hop

In attempts to gain more and more anti-squat, it soon becomes apparent that the effective swing-arm length of the rear suspension will become shorter and shorter. Like most adjustments on a car, you reach a point of compromise. When the side-view swing arm of the rear suspension is too short, rear axle hop during braking becomes a problem.

When the swing arm gets short, the axle housing torque needed to resist the braking forces can raise the rear wheels off the ground. If the wheels are off the ground, the brake forces go to zero and the wheels return to the ground. This violent cycling, off and on, produces *brake hop*. Based on our experience, a swing-arm length of at least 42 inches is needed to prevent rear axle hop. As Figure 9-9 shows, this swing-arm length also restricts how much anti-squat can be had.

TYPES OF LIVE AXLE REAR SUSPENSION

Now that we've established our design requirements, let's take an indepth look at the types of live

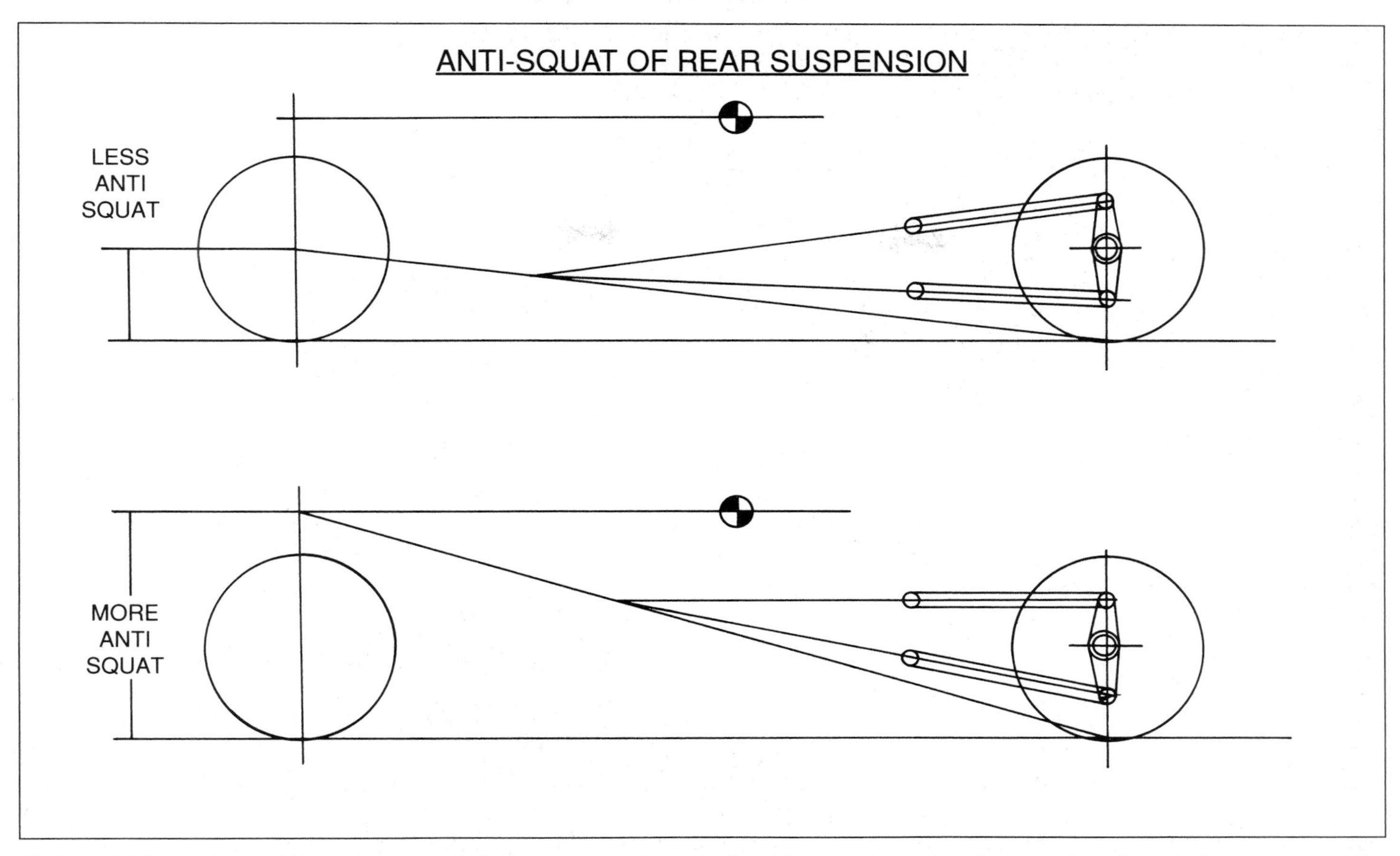

Figure 9-8. The location of the instant center of the rear control arm in the side view determines how much anti-squat the rear suspension has. More anti-squat permits the driver to apply the throttle sooner coming off a corner.

axle rear suspension to see how well they meet this criteria. Any one of the following types can be designed to provide satisfactory results. Usually, the quality of the design and installation is more important to the quality of handling than the type of rear suspension used.

Hotchkiss Drive

A Hotchkiss drive rear suspension is one where a longitudinal leaf spring is used on each side of the car to locate the rear axle (Figure 9-10). These springs also support the rear vehicle weight, so it is a very simple system. Since it is a carryover from the horse-and-buggy days, it has been well developed over the years. In 1970, Chevrolet found that placing one of the shocks ahead of the axle and placing one of them behind it would eliminate torque-induced wheel hop under acceleration and braking. This fairly recent development has allowed the Hotchkiss drive to stay relatively current. The major disadvantages of a Hotchkiss drive rear suspension are the weight and size of the leaf springs. Many cars do not have room to position the springs in their proper location and this limits their application in favor of other configurations.

Adding Spring Leaves—Since rear leaf springs are also used to transmit acceleration and braking forces, their positioning and configuration is critical. These forces cause the axle housing to twist, and in doing this the leaf springs tend to wrap up and distort. To counteract this condition, it is necessary to increase the number of leaves or the thickness of the leaves to give the spring enough stiffness to resist these forces. Unfortunately, adding leaves or making them thicker also increases the spring rate. To solve this conflict, leaves can be added only at the front of the leaf spring to provide the spring with the stiffness it needs in the front half to resist the axle torque reaction, while not increasing the spring rate as much as if a full leaf was used. Most leaf springs on high performance cars have extra leaves only on the front of the spring for these reasons.

Spring Eye Height—The height of the front spring eye primarily determines how much anti-squat the

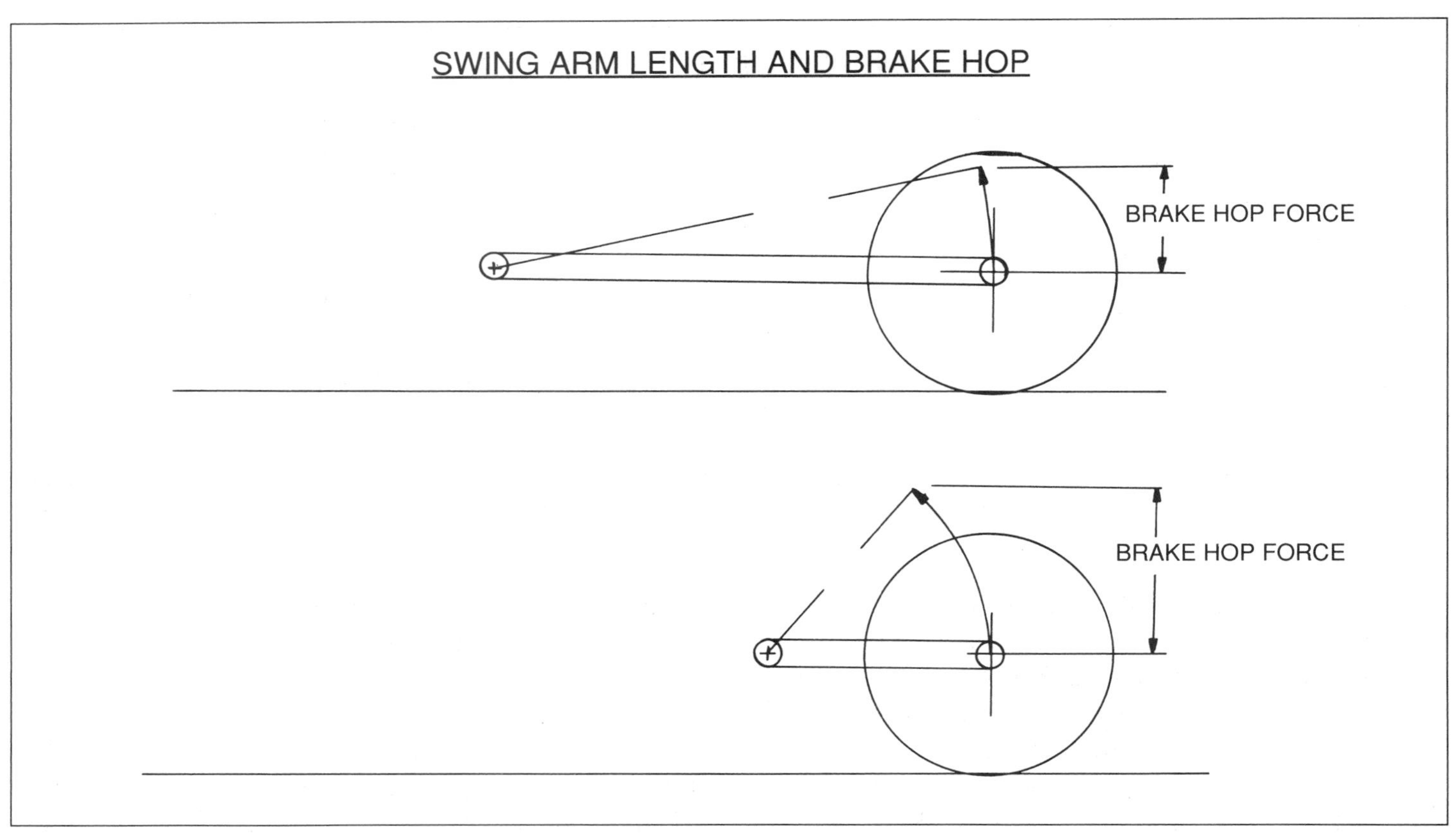

Figure 9-9. Shortening the swing arm increases the force to pull the rear wheel off the ground for a given axle torque during braking.

rear suspension will have. The higher the spring eye, the more anti-squat. Generally, heights between 10 and 15 inches above ground give the best results. If you get the front spring eye too far off the ground, you will get brake hop, so there are limits. If you are building a new chassis, make the front spring eye height adjustable, so you can tune your chassis in this area. The height of the rear spring eye also affects anti-squat, but to a much lesser extent. Adjustable-length rear shackles usually allow for fine-tuning the anti-squat and the ride height. Because of the stiffness of the leaf springs in the lateral direction, it is unnecessary to use a Panhard bar or a Watt's link to locate the axle sideways in the car.

Shock Mounting—Mounting the shocks on a rear leaf spring suspension requires special consideration. If one shock is placed in front of the axle, and one shock is placed behind the axle, their dampening effects do a good job of controlling the violent torsional rotation of the axle during power hop and brake hop. This configuration is so successful that it eliminates most of these problems without any other devices. There is no preferred way of mounting the shocks, so either the left or right shock may be mounted behind the axle. Under extreme conditions, traction bars and/or telescoping brake rods can be used to provide additional torsional control of the axle on a Hotchkiss drive rear suspension.

Leaf Spring Bushings—The design and materials used for leaf spring bushings is an important consideration when selecting them for your application. For normal use, the stock-type rubber bushings are best. High-performance driving can be improved with a spherical bearing in the front spring eye. Under no circumstances should you use urethane bushings on the front spring eye, as they will bind up and prevent the body from rolling in relation to the axle.

Link Type

With a link-type rear suspension, the live axle is located both longitudinally and laterally with link-type members. The arrangement of these links can be used to provide different characteristics, so there are a number of possibilities.

Four-Link—A four-link rear suspension uses four longitudinal links to locate the axle fore and aft, and to control the axle torque loads due to acceleration

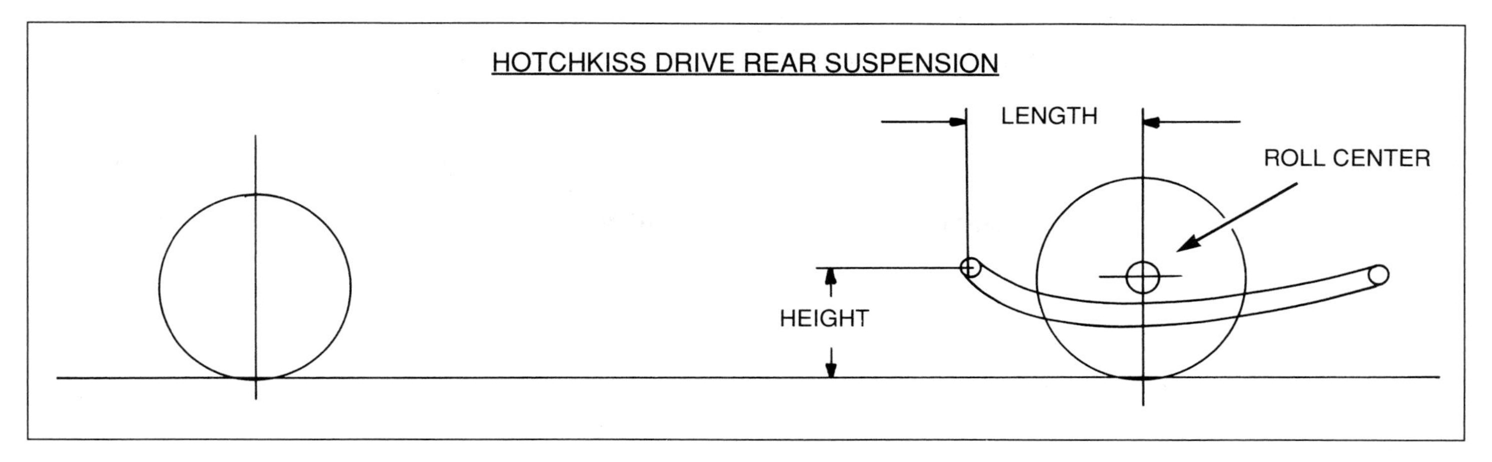

Figure 9-10. A Hotchkiss drive rear suspension utilizes a leaf spring at each end of the axle housing to locate the housing and to provide the springing medium. There is no need for any other members to locate the axle sideways in the car, because of the stiffness of the springs in the lateral direction.

and braking (Figure 9-11). Lateral location of the rear axle is controlled by the use of a Panhard bar, a Watt's link or similar device. Four-link systems work best when the links can be long and when a minimum of anti-squat is required. It is difficult to get roll understeer with an anti-squat geometry on a four-link rear suspension. In order to allow the body to roll in relation to the rear axle, the links on each side of the car must remain parallel with each other. This severely limits the geometric variations possible, so I recommend using the other types of link rear suspension.

Three-Link—A three-link rear suspension uses only three links to locate the axle longitudinally in the car (Figure 9-12). With the three-link system, there is more opportunity to get the optimum roll steer and anti-squat characteristics without restricting the roll angle between the axle and the car. Many short-track race cars use a three-link rear suspension, so the system obviously works very well for high-performance applications.

The height of the brackets that attach the links to the rear axle housing are usually about 5 inches from the centerline of the axle shaft. You can make them shorter, but then the forces in the links get higher and the Heim joints will wear quicker. If you make them longer, packaging becomes a problem. Any dimension between 3 in. and 7 in. will probably work okay.

With a normal three-link rear suspension, it is possible to get over 100% anti-squat. Unfortunately, this amount of anti-squat is difficult to get with roll understeer, because the linkage arrangements are different. A high degree of anti-squat with roll understeer is possible, but it results in a very short side-view swing arm length. Remember, a short swing arm causes brake hop when it gets too short in length.

In order to get roll understeer with a three-link rear suspension, you usually need to place the front of the lower arms lower than the rear of the arms. Unfortunately, this conflicts with the anti-squat requirements. The compromise is usually to mount

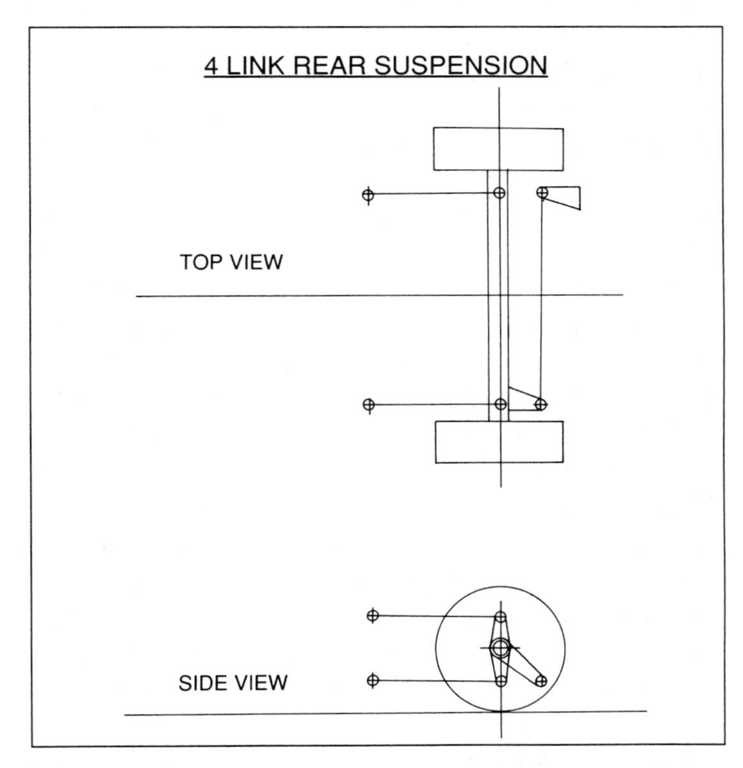

Figure 9-11. A 4-link rear suspension uses four control arms to locate the axle housing fore-and-aft in the car. Since these four arms provide no lateral control, a Panhard bar or similar device is needed.

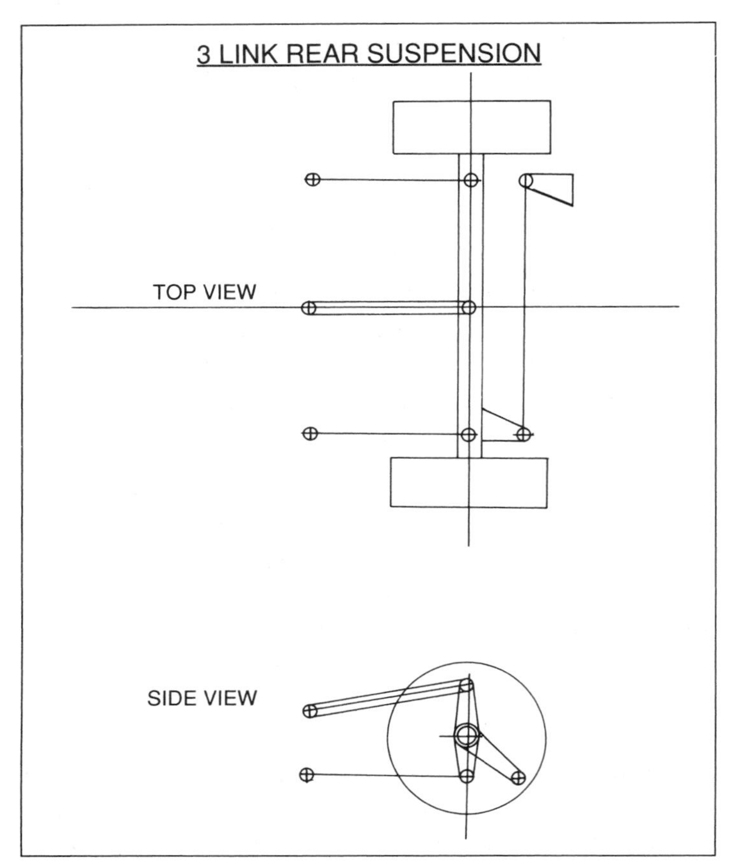

Figure 9-12. A 3-link rear suspension is similar to a 4-link, except it is a bit more simple. It has the advantage of allowing more roll angle between the axle and the chassis while still providing some anti-squat characteristics.

the lower arms parallel to the ground for a minimum of roll steer and then set the upper arm angle to provide the best anti-squat with a long enough side-view swing arm.

For low- and medium-powered cars, this compromise can be made to work well. For high-powered cars, or for the optimum in performance handling, it is necessary to decouple the three-link system so that each of the elements can be optimized without affecting the other elements. This decoupling action is what makes the torque-arm rear suspension attractive, which is discussed next.

Angled Arms—One disadvantage with the three-link and four-link rear suspensions is the need for a device to control the lateral location of the axle in the car. A Panhard bar or a Watt's link is usually used for this purpose. A Panhard bar is simpler and therefore more commonly used. But the bracketing needed to rigidly mount a Panhard bar to the axle and to the chassis is heavy and takes up space.

The need for extra links to locate the rear axle laterally in the car can be eliminated by angling two of the links on a four-link system. This is a very simple solution to the problem, and if these links are arranged correctly, it is possible to get both good roll steer and good anti-squat characteristics. The best arrangement of the angled arms is called a *Satchell link* rear suspension. This configuration places the angled arms below the axle with their front pivots pointing toward the center of the car. The advantages of this arrangement are low roll center, good anti-squat, good support of the axle housing ends, and little need for extra frame bracing. The system is also compact so it can fit many types of cars. Of all the link-type rear suspensions, the Satchell link has the most benefits (Figure 9-13).

Torque Arm

A torque-arm rear suspension uses a long arm to absorb the rear axle torque reactions (Figure 9-14). Other suspension members are needed to locate the rear axle laterally and longitudinally, as well as to

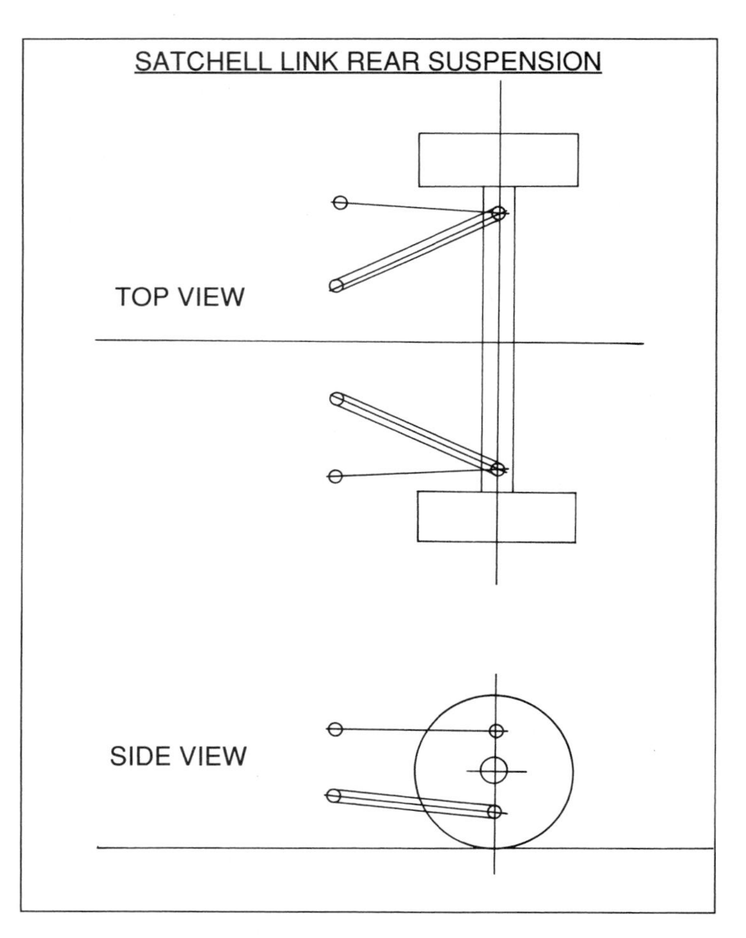

Figure 9-13. A Satchell link rear suspension is a variation of the 4-link and the angled arm suspensions in that it only uses four links. The advantages of the Satchell link are that it provides considerable anti-squat with roll understeer and a low roll center. It also supports the ends of the axle housing very well.

provide the springing medium. Torque-arm rear suspensions were used on Fords in the 30's and 40's, on Chevy trucks in the 50's and 60's, and most recently on Chevrolet Vegas and Camaros. Torque-arm rear suspensions can be packaged efficiently and when correctly designed, they can be tuned to provide good overall performance. The interaction of the torque arm, the rear suspension and the chassis are more complex than it seems, so it is sometimes difficult to get optimum performance without extensive testing and development.

The same design factors that are important to a three-link are important to a torque-arm rear suspension:

1. Lateral location must be maintained with a low roll center to provide consistent and predictable handling. You can use a Panhard bar or a Watt's link.

2. Axle housing support is again important, but since the axle torque reaction is absorbed by the torque arm, the two lower control arms can be raised to axle centerline height, because their only requirement is to transmit the driving forces fore-and-aft into the chassis. Moving these links up prevents the driving forces from twisting the axle housing. Since the outboard links do not need to absorb the axle torque reaction, their position is not as critical to the anti-squat function. This means that they can be arranged to provide some roll understeer without the usual compromises.

3. As I mentioned previously, the real advantage of a torque arm is its ability to absorb the axle torque reaction as an independent component. A torque arm can be a rigid member mounted to the chassis by a separate spring-shock unit, or it can be a flexible arm such as the type used on Reese equalizer trail hitches.

Reese Bar—The use of a Reese bar spring torque arm allows the loads to be applied to the chassis in a smooth and progressive manner. Other types of springs, like fiberglass, can also be used. The stiffness of the spring determines how much axle housing rotation there will be for any given torque. The action of a torque arm under acceleration is similar to the traction bars used in drag racing. The shorter the torque arm, the more anti-squat it will produce. Making the arm too short will obviously cause severe angles and limit the free motion of the suspension in the vertical direction. From a practical standpoint, you should probably not run a torque arm much shorter than about 30 inches. Raising the torque arm in the car will not cause a significant change in the amount of anti-squat.

Decoupling—When a torque arm is connected directly to the chassis, the braking torque of the rear

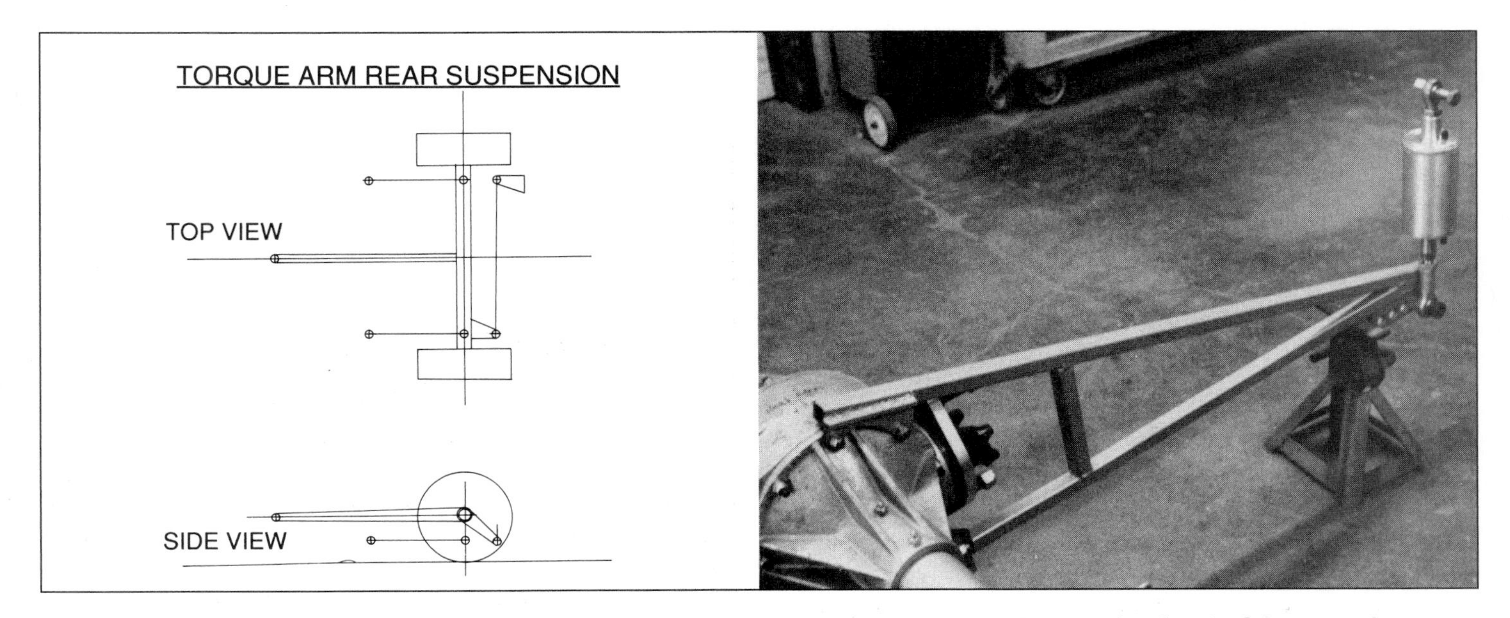

Figure 9-14. A torque arm rear suspension uses a separate member to control axle torques relative to the chassis. Other control arms are needed to locate the axle housing in the car.

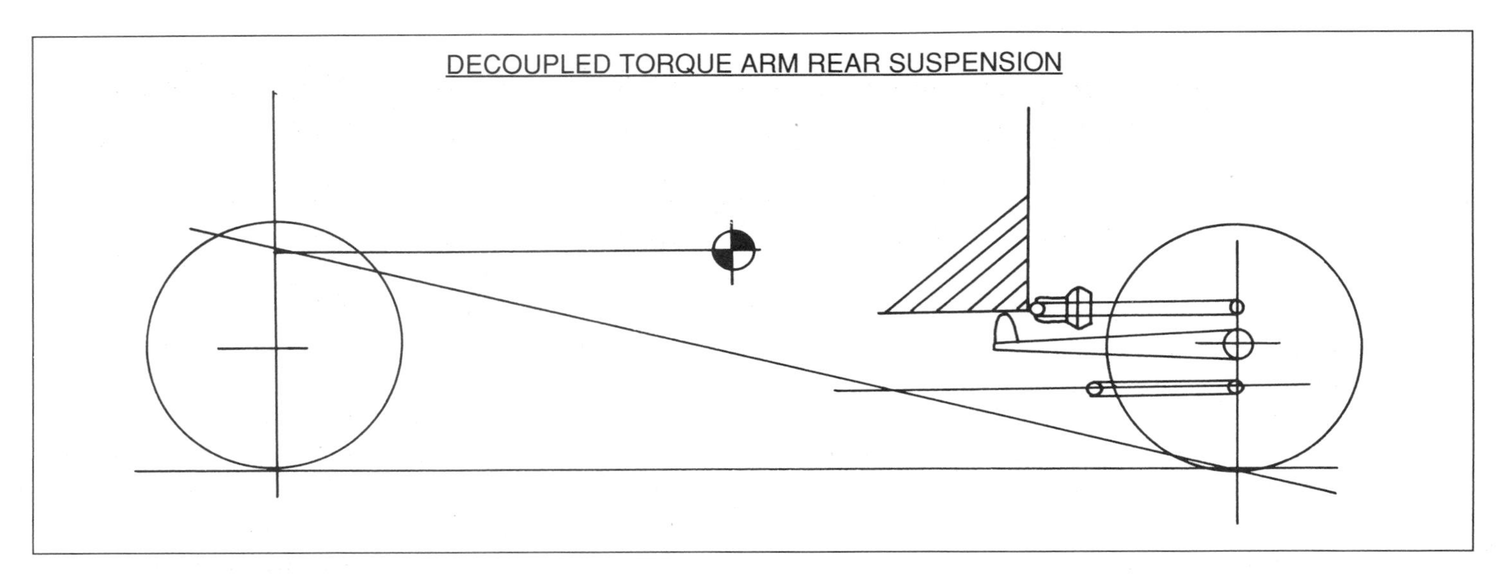

Figure 9-15. Decoupling the torque arm can help to eliminate brake hop. The arm is mounted so it contacts the frame during acceleration, transmitting the axle torque to the frame, yet it is free during braking. By separating the braking and accelerating absorbing functions, it is possible to optimize each without compromising the other.

axle can cause the rear axle to lift off the ground because it can pivot about the torque arm drop link. This action is the same as when the side-view swing arm is too short, so brake hop will result.

A solution to the brake hop problem is to decouple the torque arm so that the axle torque from acceleration is absorbed by the torque arm, and the axle torque from braking is absorbed by a different member. One way to do this is to let the front of the torque arm contact the frame only in the "up" position. This allows it to be free from the frame during braking while allowing it to transmit the axle torque reaction to the frame during acceleration. A rubber bumper can be used to soften the impact when the torque arm contacts the frame.

Because the torque arm is disconnected under braking, another suspension member will be needed to absorb the axle torque during braking to eliminate brake hop. A good way to solve this is with a *telescoping upper link*. This lets you optimize the torque arm configuration without having to compromise it with brake hop characteristics. A telescoping upper link should be arranged so it bottoms out when the axle rotates because of the braking torque. When this happens, the upper link acts like a solid member and together with the lower links, establishes parallel linkages with a very long side-view swing-arm length. This long swing-arm length eliminates brake hop, so you have the best of both conditions. The torque arm absorbs the axle torque reaction during acceleration and its short length provides a lot of anti-squat. The telescoping link absorbs the axle torque reaction during braking and its effective linkage is long enough to prevent brake hop.

Establishing the optimum spring rates of the bushings, and finding the correct amount of preload for the total system, might take some testing. If you arrange the links as shown in Figure 9-15, you will be pretty close to achieving the right setup. Figure 9-16 shows a decoupled torque arm with a telescoping upper link mounted on a late-model Camaro axle. ■

Figure 9-16. A telescoping upper link can be used with a decoupled torque arm to absorb the axle torque reaction during braking and its effective linkage is long enough to prevent brake hop.

INDEPENDENT REAR SUSPENSION DESIGN 10

An independent rear suspension (IRS) design is one where each rear wheel is not connected to the other wheel. On a rear-wheel-drive car, this means that the differential is mounted solidly to the chassis with some type of driveshaft going to each wheel. (Figure 10-1).

Advantages

The main advantage of an independent rear suspension is that it provides for a smoother ride, especially over rough pavement. This advantage, however, does not occur automatically. This was demonstrated by the 1984 Corvettes, which used a very stiff rear spring and therefore had very poor ride characteristics, even though they had independent rear suspension. Most of the smoother ride, and potentially better road-holding, advantages of an independent rear suspension come from having the differential mounted to the chassis. This reduction in unsprung weight allows each rear wheel to follow a rough road surface better, because the differential does not have to bounce as far with the wheels. If a wheel can follow the road more of the time, it will maintain its traction more of the time, and this will increase its road-holding capability. The other advantage of an independent rear suspension is that it can take up less room in the car.

Disadvantages

The main disadvantage for an independent rear suspension is that it is more complex, so it costs more. The other disadvantage is that because of this complexity, it is more difficult to design one correctly. On production cars, like the Corvette, the designers need to simplify the system in order to save cost. This often results in a compromised design which does not provide optimum geometry or deflection characteristics.

Another major disadvantage of an independent rear suspension is its ability to deliver power to the ground. The basic geometry of an IRS is such

Independent rear suspensions offer the advantage of a smoother ride and better traction over rough pavement. However, because of its complexity, it is difficult to design one properly, as some model-year Corvettes have demonstrated. Mandates to control costs have forced Corvette engineers in the past to compromise the design, which led to problems like poor camber control and deflections. However, '84 and later Corvettes have been steadily improved, especially those from 1988 to present.

that only small amounts of anti-squat can be used, so high-powered cars without a rear weight bias will have problems getting optimum performance and handling.

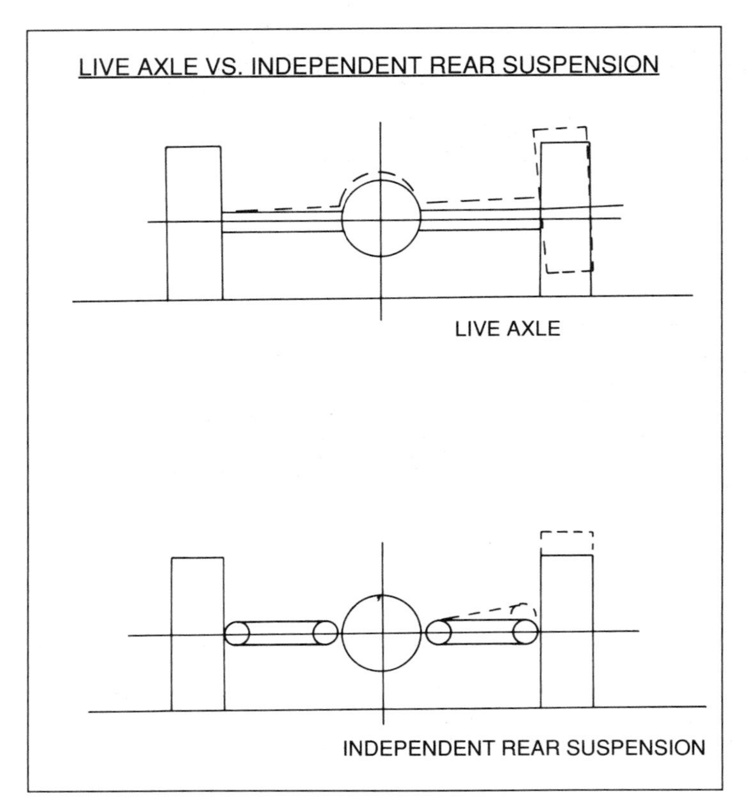

Figure 10-1. An independent rear suspension has each rear wheel controlled separately from the other. This arrangement requires jointed drive axles to transmit the power from the differential to the wheels.

DESIGN REQUIREMENTS

The design requirements for an IRS are similar to those for a live axle rear suspension. Basically, the rear wheels need to be kept pointing in the right direction at all times, and the camber and roll center need to be positioned as consistently as possible. To do this, you will need to make the components rigid and secure so they don't move under loading, and you will have to use suspension geometry that produces smooth and consistent wheel motions. All of these factors are based on attaching the rear suspension to a frame that is torsionally rigid. Without a rigid frame, the precision of the suspension is obliterated by chassis flex. Frame design and build are covered in Chapters 12 & 13.

The following aspects should be considered when designing an independent rear suspension.

Anti-Squat

Our previous analysis of live axle rear suspensions in Chapter 9 has shown that one of their advantages is that they can be designed to provide a fair amount of anti-squat, which can be used to get a lot of power to the ground. This is a real advantage during high performance driving, so you should try to incorporate this feature into the ideal independent rear suspension. Unfortunately, this is a very difficult thing to do, since the differential torque reaction is not transmitted through the suspension members on an independent rear suspension. By properly locating the rear suspension members, it is possible to get some anti-squat, but a value of about 25% is the practical limit (Figure 10-2). This compares to a well-designed live axle rear suspension, which often has over 100% anti-squat.

Because independent rear suspension systems do not have as much anti-squat, they are not as good as a live axle in getting the power to the ground on high performance cars. This problem is not too bad with a rear-engined car, because of the rear weight bias. On front-engined cars however, it is a distinct disadvantage.

Rear Suspension Geometry

The type of suspension geometry required for an independent rear suspension system is not too different from that required for an independent front suspension. On a front suspension, roll understeer is obtained by having the wheels toe-out as they go up into jounce. On a rear suspension, roll understeer is obtained by having the wheels toe-in as they go up into jounce. As shown in Chapter 4, understeer at the front is accomplished by placing the steering gear ahead of the axle because of deflections of control arm rubber bushings. With independent rear suspension, deflection understeer is accomplished by placing the steering rods behind the axle.

I believe that it is good design to use a long swing-arm length and a low roll center on an independent rear suspension, just as on an independent front suspension. These factors result in smooth and consistent wheel motions which give the driver confidence and better control. Roll angles can be controlled with stabilizer bars, front and rear, so there isn't the need for the roll resistance that a high roll center gives. Camber gain should be positive, but

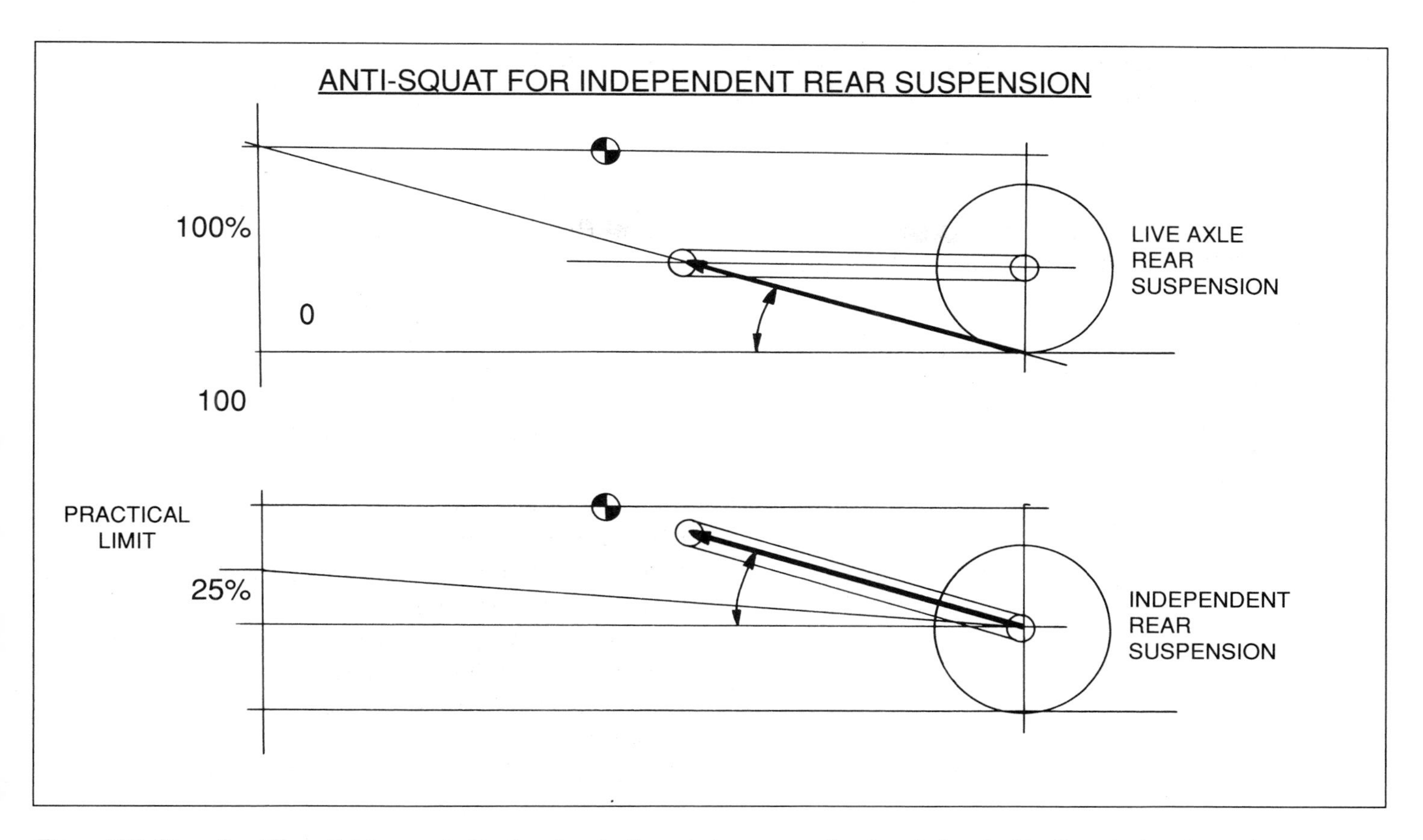

Figure 10-2. Since the differential is mounted to the chassis, the axle torque reaction is not absorbed by the axle housing and rear suspension. This has the effect of reducing the anti-squat of the rear suspension, because the only vertical force available for this purpose comes from the angle, if any, of the wheel hub control arms, which do not see any torque reaction during acceleration.

again the stabilizer bars will limit the roll angle so aggressive camber gain is not needed or desired.

Springs & Shocks—Spring and shock rates, and mounting locations, on an independent rear suspension are similar in design as a front IRS. The major difference is the need to offset the springs and/or shocks to clear the drive axles at the rear. This usually requires offset mounting brackets, rocker arms or mounting the springs/shocks above the rear knuckles. The more complicated the system, the more it will cost, and the greater will be the need for additional maintenance. From a handling standpoint, it doesn't matter so long as the wheel rates and wheel travel are the same.

TYPES OF IRS

To help demonstrate these design requirements, I will do a brief analysis of three popular types of independent rear suspension.

Corvette

The Corvette IRS design suffers from a variety of geometry and deflection problems.

The early Corvettes up through 1987 had a roll center height that was too high. This condition resulted in a jacking problem, wherein the cornering force from the outside tire would cause the rear of the car to raise up. This increase in rear height would change the camber of the outside tire, so it lost cornering force. The result is a car that must be set up with an excess of understeer in order to keep the average driver from spinning out during hard cornering.

Even with perfect suspension geometry, a car can exhibit inconsistent handling behavior if there is excessive deflection of any of the components. The early Corvette design provides many places where deflections can result in poor handling. Some of these are:

Camber Control—A lower camber rod is used to maintain the camber angle of the rear wheels. On

Figure 10-3. The early Corvette IRS design suffers from a variety of geometry and deflection problems. Chief among them was a high roll center height, which caused jacking and a loss of cornering force. Photo by Michael Lutfy.

paper, this design looks good but since this rod uses rubber bushings, its effective length can change under cornering loads. This change in length produces positive camber on the outside wheel, which also reduces its cornering power. Reducing the cornering power of the outside tire during hard cornering, when you need it most, is not conducive to good handling. A similar problem exists where the drive axle attaches to the differential. Since the drive axle is actually the upper control arm, any wear or looseness of the differential side-gears also results in positive camber during hard cornering. Even more deflection is introduced to the system where the differential mounts to the chassis. Rubber mounts are needed to isolate the axle gear noise, but they also allow the differential to move in relation to the chassis during hard cornering, because all the lateral loads are fed into it.

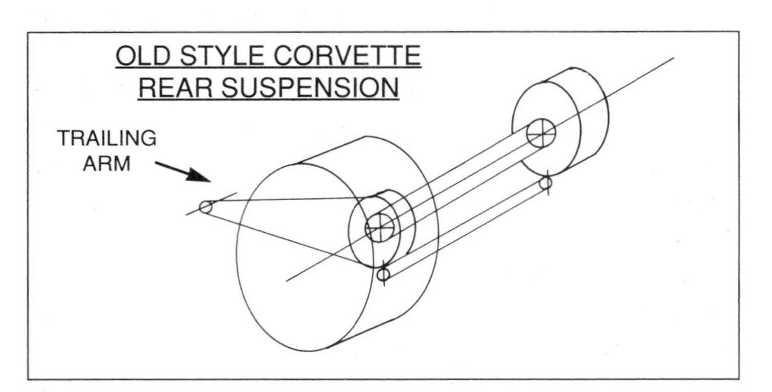

Figure 10-4. The first Corvette independent rear suspensions were designed with an eye on cost-saving. The simplicity of the system required some compromises in camber and toe control that resulted in peculiar handling under certain circumstances.

Toe Control—The '63-'82 Corvettes used a single trailing arm on each side to locate the axle hub fore and aft on the car. These arms also provided the rear-wheel toe control. Although these trailing arms are well located by design, the rubber bushings used where they attached to the frame allow toe deflection when the power or the brakes are applied (Figure 10-4). These changes in direction do not result in consistent handling.

The '84-and-later Corvettes use two trailing arms on each side together with a rear tie rod to effectively eliminate the torque steer problem of the earlier design (Figure 10-5).

Jaguar

As the Jaguar independent rear suspension design uses the drive axle for the upper arm and a lower arm to control the camber, like the Corvette, it

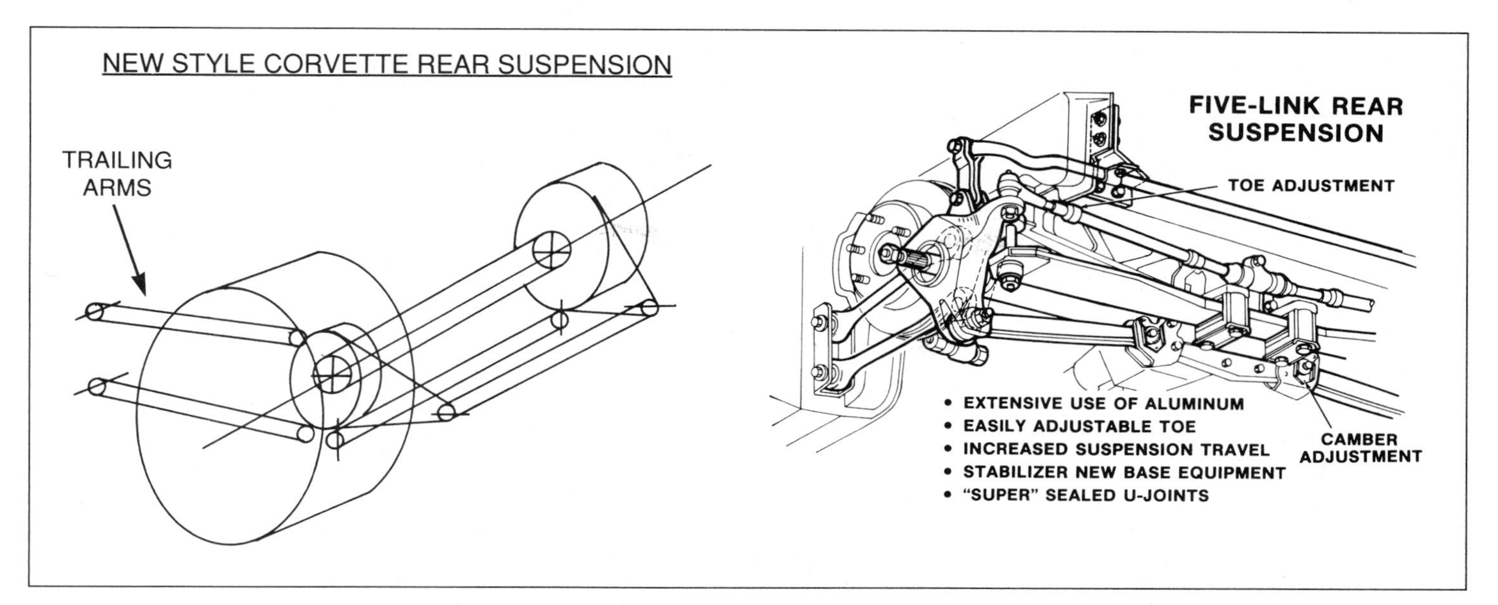

Figure 10-5. Experience with the old system and more advanced knowledge of suspensions in general, resulted in an improved Corvette independent rear suspension system beginning with models after 1984. There are still some flaws in the design, but its use is satisfactory for most applications.

suffers from the same camber control problems. In addition, the Jaguar also has deflection steer problems because the toe control is achieved through the same lower arm. Any looseness or wear on these lower control arm bushings will result in loss of toe control. Just like the earlier Corvettes, Jaguars experience torque steer problems. These problems become much worse when high-powered engines and sticky tires are used (Figure 10-6).

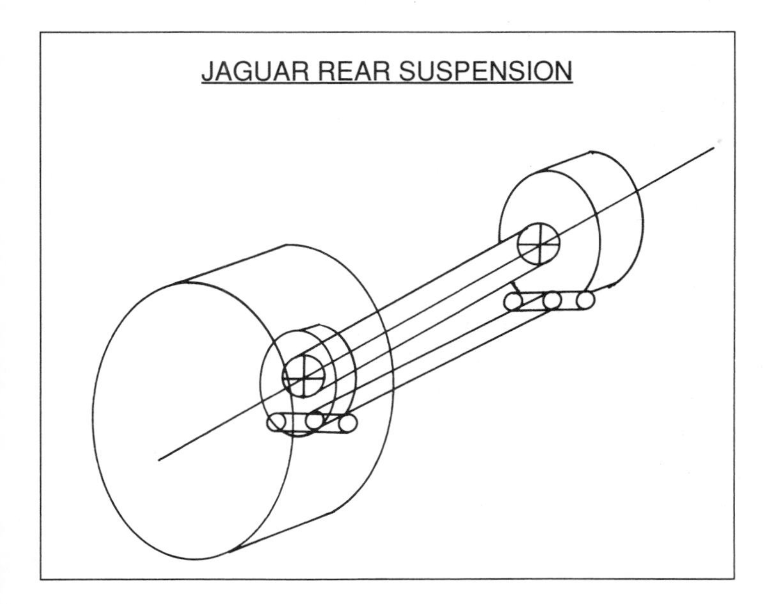

Figure 10-6. The Jaguar type of independent rear suspension is simplicity itself. On paper it looks good, but in actual use it suffers from deflection and loose connection problems. It is not the preferred system for use on high-powered cars.

Also like the Corvette, the Jaguar feeds most of the suspension loads into the differential. This means that any looseness of the side gears can result in loss of camber control. The rubber mounts between the differential and the frame can also deflect, allowing the whole rear suspension to move around in the car.

Both the Corvette and Jaguar independent rear suspension systems are adequate for normal highway driving. High performance driving, however, requires that the suspensions carry very high cornering loads, and these inexpensive independent rear suspension systems do not perform under these conditions as well as they need too.

Formula Cars

The types of independent rear suspension used on formula racing cars are examples of the correct way to design the system (Figure 10-7). Obviously, these designs are not built to the lowest possible cost. Some of the design features of these systems that give precise control of the rear-wheel camber and toe are:

No Rubber Bushings—Formula cars use Heim joints instead of rubber bushings, because they do not allow unwanted deflections (Figure 10-8). Rubber bushings are used on street-driven cars, because Heim joints tend to be noisier. If rubber bushings must be used, they should be arranged so their deflections result in a minimal change in handling.

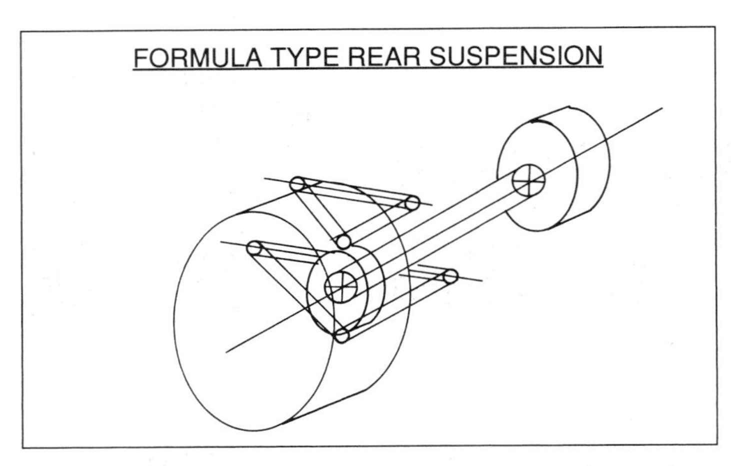

Figure 10-7. The type of independent rear suspension used on Formula racing cars is very similar to a normal independent front suspension. The use of rigid control arms accurately positions the knuckles and wheel for consistent handling regardless of the power level or cornering forces involved.

Rigid Members—Most formula cars use two members to effectively triangulate the location of the top and the bottom of the rear wheel hubs. This accurate and precise location of the rear wheels assures that the design geometry is actually achieved on the car. Formula cars also use rigidly built knuckles and hubs to prevent them from bending under high load conditions.

Suspension Loads—Formula cars use upper and lower control arms to transmit the cornering loads to the chassis. A *slip joint* is used in the drive axles to permit the changing length. On those designs where the suspension loads do go through the differential, the differential is solidly mounted to the chassis, so it cannot move around during cornering, braking or acceleration.

SUMMARY

A well-designed independent rear suspension can be made to handle as well as a well-designed live axle rear suspension. The only basic limitation is that the IRS will not have as much anti-squat, so it won't get the power to the ground as well. Under high horsepower and high performance driving conditions, this can be a disadvantage.

The major advantages of an independent rear suspension are the potential for a smoother ride and better adhesion over rough pavement. Under these conditions, the independent rear suspension can be better if the system is correctly designed. This potential advantage must also be evaluated with regard to the added cost required. ■

Figure 10-8. Formula cars use Heim joints instead of rubber bushings to eliminate deflection. Accurate and precise location of the rear wheels is assured by the use of two members to triangulate the top and bottom of the rear wheel hubs, resulting in optimal design geometry. Photo by Michael Lutfy.

BUILDING REAR SUSPENSIONS 11

LIVE AXLE REAR SUSPENSIONS

Even the best rear suspension design will not work well unless it is built to the highest standards. Like most aspects of car design and building, there is no best way. Every car builder has his own solutions to problems, and you can learn from all of them.

As an example of one way to build a live axle rear suspension, here are some of the problems and solutions I encountered in designing and building a short-track race car. On this project, I kept our options open on the location of the links so I could use a test program to optimize the package. The criteria for selecting components for a rear suspension are the same as for any other part of the car. You want strength and rigidity along with as light a weight as possible. Proven durability is also important, and to some extent, cost must also be considered. Since this building example is for a short-track race car, the choice of components reflects the intended use.

Rear Axle Housing

The major component of the rear suspension is the axle itself. Since this is a major investment, it is wise to make the selection carefully. The two most popular types of axles used for short-track cars are the Ford 9-inch (Figure 11-1) and the Quick-Change.

Ford 9-Inch—The Ford 9-inch axle is strong enough for most racing applications and it is fairly light and compact. This translates into reasonable unsprung weight and *rotational inertia*. There are plenty of gear ratios available, but it is necessary to change the whole third member to make a change. The Ford 9-inch axle uses a *hypoid* gearset with considerable offset. This type of gearset has a fair amount of rubbing friction, which causes heat. An axle cooler is required with a 9-inch axle if it is to run for any long period of time, in order to dissipate the heat. The amount of heat is an indication of the friction involved, and of course, friction reduces horsepower.

No matter how good your rear suspension design looks on paper, it won't work as it should unless it is built to the highest standard. There are few examples better than a NASCAR Winston Cup car.

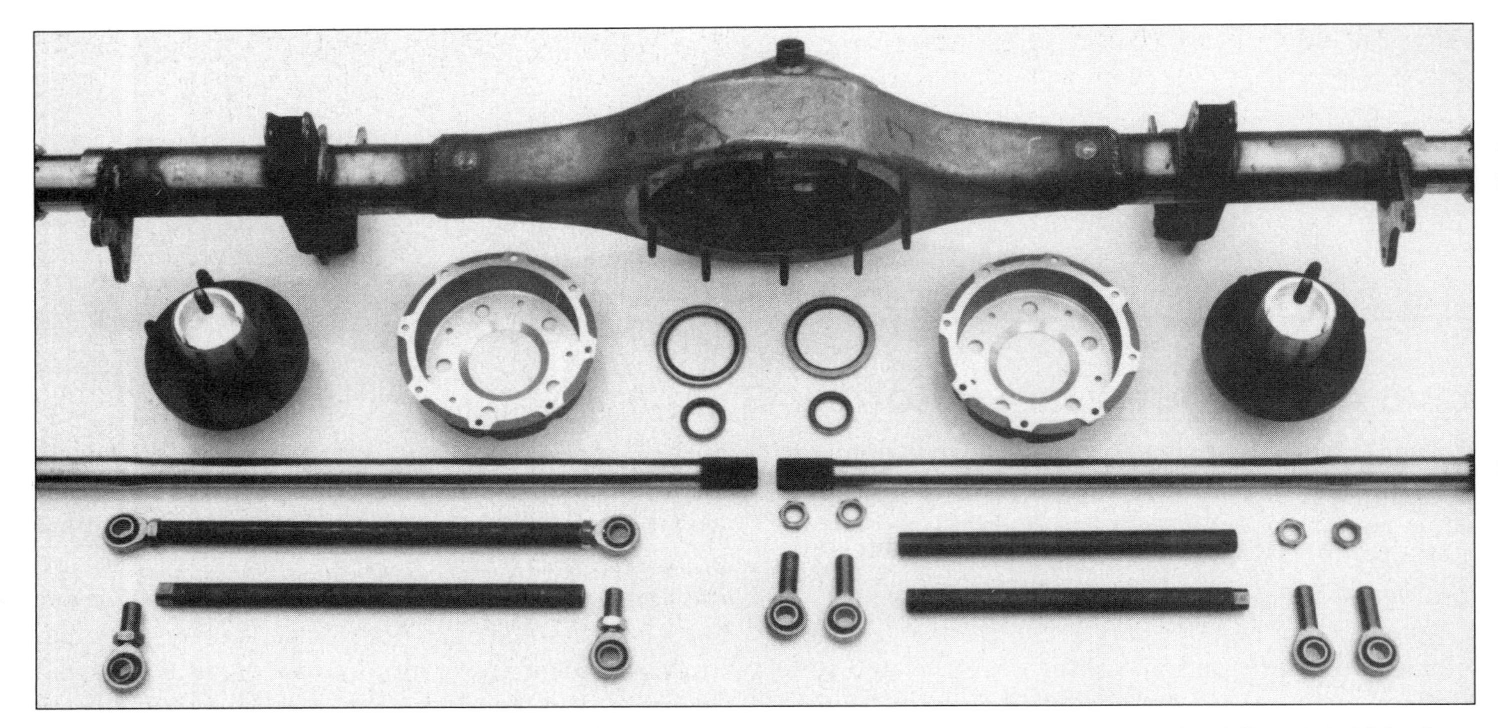

Figure 11-1. The Ford 9-inch axle is popular for race cars and specialty cars because it features interchangeable differentials. This allows a way to quickly change gear ratios and/or differential types. The 9.00 inch ring gear is strong enough for almost any power level.

Quick Change—A Quick-Change axle uses Spiral-Bevel gears, so there is less friction generated at the ring-and-pinion. However, the added transfer gears add some friction (about 1% loss in efficiency), which must be considered as part of the total. The fact that most Quick-Change rear axles don't use a cooler is an indication that they are more efficient from a friction standpoint. Quick-Change axles are also superior from the standpoint of gear selection and ease of changing ratios. It is also more economical to have a selection of gear changes, because only the transfer gears are needed, not entire third members.

Axle Lateral Locating Systems

Previously I discussed various means of locating the rear axle in the car laterally. I built our project car with provisions for both a Panhard bar and a Watt's link. This selection allowed for testing each to determine if one had a handling advantage. From a design standpoint, the handling should be the same for both of them since I kept the same roll center height.

The Watt's link requires less frame structure than the Panhard bar, but it is more complex, so neither of these systems offers a big design advantage over the other.

As can be seen from Figure 11-2, all of the brackets used to mount these various axle locating devices have been designed to provide rigid structures. If the brackets bend, or if the frame bends, the real handling effects will not be known. All of the frame and axle brackets have been made with a selection of attaching holes, so the anti-squat and roll center heights for each configuration can be adjusted.

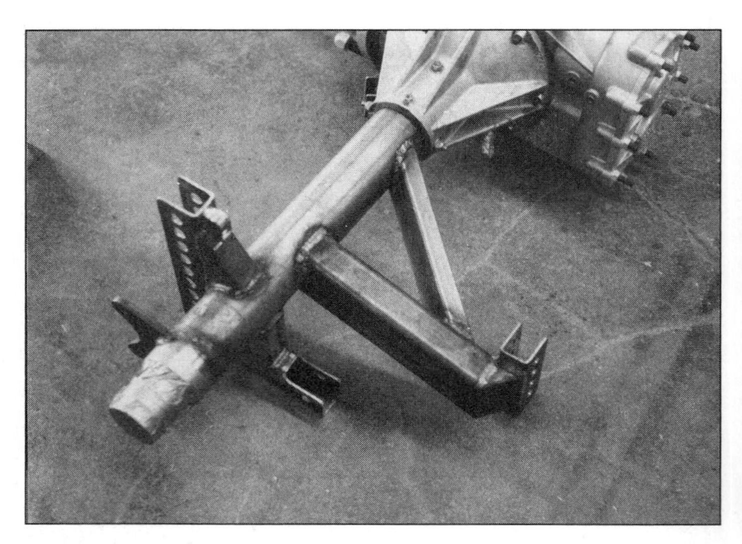

Figure 11-2. The only way the rear axle movement can be kept to the control arm geometry is if the axle brackets are rigid enough to resist bending. The loads are very high, so the brackets must be heavy-duty.

Axle Links to Absorb Driving Forces

The project car was built so both three-link and four-link systems could be tested to see if there were any handling differences. With adjustable control arm brackets on the frame and on the axle, a wide variety of anti-squat and roll steer values could be tested. With either a three-link or a four-link system, the anti-squat forces are developed by the arrangement of the trailing links.

Torque Arm—Another system for absorbing axle torque and driving force is to use a torque arm. With this system, the driving forces are absorbed by a single link at each side and the axle torque is absorbed by a centrally mounted arm. The advantage of this system is that it is possible to have more anti-squat without getting brake hop, because the axle torque from acceleration and braking are controlled by two separate systems. I built the car so I could also test a torque arm rear suspension. Torque arm rear suspensions have similar anti-squat characteristics to leaf spring rear suspensions, so there is the possibility of getting more traction for coming off the corners with this system.

Rear Axle Camber & Toe-In

Like the front tires, camber and toe-in can be used to help the rear tires perform to their maximum potential. A tire delivers its maximum cornering power when it has a slight amount of negative camber in reference to the ground. Some of this is the result of the bending of the wheels, hubs, axles and control arms. Regardless, you can improve the total tire cornering forces of your car if the rear tires run at a slight angle of negative camber. On circle tracks, where the cars only turn left, negative camber on the right rear tire and positive camber on the left rear tire are desired.

Bending the Axle—Rear tire stagger provides this condition, but it can also be done with a solid beam axle by bending the axle tubes so they are at an angle to the ground. The axle splines will accept one degree of misalignment, so this is a good objective. I bent the axle tubes by heating the side I wanted to shrink to make the bend. On the right side, this is the top of the tube, and on the left side, it is the bottom of the tube. The amount of bend is dependent on how long a section of the tube is heated. Heating a spot of the tube 6.0 inches long until it is red-hot will usually heat-shrink it for a 1/2 degree bend. When you heat the tube, support it so its weight helps to cause the bend. After heating, let the tube air-cool. Obviously, you must remove all the oil from the axle before you bend it.

Figure 11-3. The rear wheel alignment is as important as the front wheel alignment. One way to change the camber or toe on an axle housing is to bend it by heat-shrinking. When a spot on the housing is heated, it will expand; when it cools, it will contract. Since it will contract more on cooling than it expands on heating, this process can be used to bend the housing in the desired direction. Negative camber results when the top of the housing is heated and cooled. Toe-in can be achieved when the front side of the housing is heated and cooled.

On a road-racing or high performance street car, both sides of the axle housing are bent for negative camber (Figure 11-3).

The same heat-shrink bending techniques can also be used to build in some rear axle toe-in. Toe-in at the rear axle is a stabilizing effect, so it is a good idea to have a slight amount (about 1/16 inch). Toe-out on the rear axle will make the car oversteer under acceleration, so you want to be sure you don't have this condition.

Because a solid rear axle can flex, it will often toe-out under braking and toe-in under acceleration. This is normally a good situation, because it makes the car want to turn-in when entering a corner while making it want to push when exiting. As long as there is some rear toe-in under acceleration, the car will be stable. If, however, there is some rear toe-out under acceleration, the car will be very loose, and the driver won't be able to get the power down. Cars that use a Ford 9-inch often have this problem, because the front face of the housing is so weak. I have used adjustable tension rods on our car to keep the axle housing from ever bending in the toe-out direction.

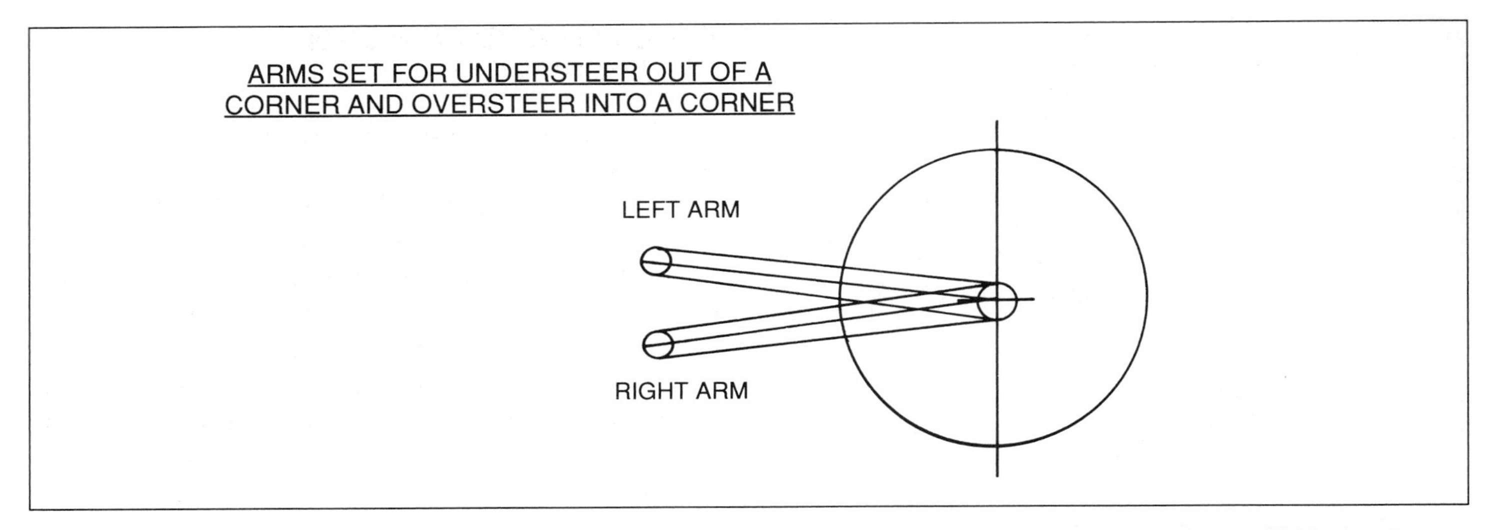

Figure 11-4. A car has a natural tendency to understeer coming into a corner and then to power oversteer coming out. If this situation can be reversed, it will make the car faster around a track. One way to do this on a circle track car is to attach the rear control arms to the axle so that when the nose goes down during braking, the rear arms turn the rear axle to steer the car to the left. Under acceleration the opposite is needed to steer the car to the right. Arms arranged as shown, can provide this feature. Note that this will not help a road-race car

Rear Axle Steering On Short Tracks

Some oval track racers locate the rear axle in the car so it always tends to make the car turn left. This is done by using longer control arms on the right rear than are used on the left rear. Obviously, quarter-mile tracks need more angle than half-mile tracks, but the effect is the same. Locating the rear axle in the car at an angle will obviously make the car want to turn left, so turn-in at turn entry is improved. Turn exit, however, is often compromised because the car wants to keep turning as the power is applied. You need to adjust the amount of angle to optimize your car's handling.

A better solution to the rear axle steer problem would be to find a way to make the rear axle angle increase on turn entry and decrease on turn exit. This would make the car want to turn-in during braking and then understeer under power, so the throttle could be opened sooner without power oversteer. With the proper rear suspension geometry, it is possible to make the rear axle steer in this manner. Most rear suspension designs have some degree of roll steer, so it is possible to make the rear axle steer angle change as the car goes from braking to acceleration. Under braking, the rear of the car tends to rise, and under power, the rear of the car tends to drop (assuming you don't have over 100% anti-squat). If the rear suspension members caused the rear axle to steer for a tighter turn when the body raised, and a lesser turn when the body dropped, you would have a rear suspension that would help turn-in on the entrance of a turn, and at the same time help to reduce power oversteer on the exit of a turn. Suspension members like those shown in Figure 11-4 would provide this characteristic.

Another way to provide right rear toe-out at corner entry and right rear toe-in at corner exit would be to use steel control arm bushings on the left rear and rubber control arm bushings on the right rear. This configuration would make the rear axle tilt rearward under braking (toe-out at the right rear), and tilt forward under acceleration (toe-in at the right rear). Letting the right side of the axle move forward and backward within the limits of the bushing deflection can provide a noticeable amount of rear-axle steer in the direction that could improve short-track handling.

Rear Suspension Links

The easiest way to make rear suspension links is to use a tubular control arm with rod-end bearings. The use of these components simplifies making rear suspension adjustments, so I installed them at all locations. Because I wanted to minimize deflections and loose joints, I decided to use expensive aircraft-quality rod ends. These rod ends use a 5/8-in. dia. bolt but have a 3/4-in. dia. shank, so they are stronger than normal. The ball, race and housing are all made of heat-treated steel for improved strength and durability.

I used thick-wall tubing for the control arms because I didn't want any more deflection than nec-

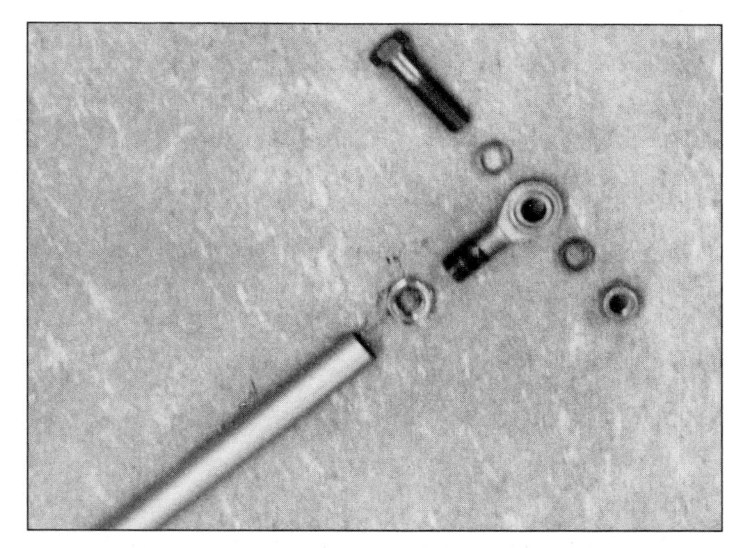

Figure 11-5. Most race cars and some high performance street cars use tubular rear suspension links fitted with rods and joints. It is necessary to use high-strength rod-end bearings and location sleeves to properly position these joints on the chassis and on the axle housing.

essary (the extra weight caused by the thick-wall tubing is at least on the rear of the car). Spacers are used at each rod end to keep it centered in the mounting bracket while allowing the maximum amount of angularity from the joint (Figure 11-5).

Rear Stabilizer Bar & Links

Most short-track cars do not use a rear stabilizer bar because weight bias is used to preload the tires for optimum handling. This system works because the car only turns left. Since road-racing cars must turn left and turn right, they have developed another system for controlling the understeer to oversteer balance. This system is based on using a rear stabilizer bar to absorb some of the roll resistance on the car onto the rear wheels. By increasing the amount of roll resistance on the rear of the car, the outside rear tire will see higher loading, so it will start to lose its cornering efficiency before the outside front tire, and the car will have less understeer. A stiffer or more effective rear stabilizer bar will increase this effect, so by being able to adjust the rear bar, you can tune the amount of oversteer you need for optimum handling. Although this technique is mainly used in road racing, it is also used on Indy cars. The use of a rear stabilizer bar can help achieve more consistent and more repeatable results. I installed a rear stabilizer bar on our project car so that I could test its effect.

INDEPENDENT REAR SUSPENSION

Building an independent rear suspension is similar to building an independent front suspension. The major difference is the need for having a drive axle going to each rear wheel. This requirement demands a more complex arrangement of the components. The rear hub must allow the drive axle to go through it, and it must be strong enough to be able to handle the drive loads involved. Running the drive axle through the hub also requires that the spring and/or shocks be mounted so that they clear the axle shaft.

Selecting the components for an independent rear suspension requires the same considerations as are used for any suspension system. You want strong and lightweight parts that are reliable and reasonably priced. The example I will be using to demonstrate building an independent rear suspension is for a kit car Cobra chassis. This is a street-driven car that occasionally runs on a road-racing course at moderate speeds. The choice of components reflects the intended use of this car.

Differential

If upper and lower control arms are used to accurately locate the rear knuckles, the choice of differential type and its mounting system are less critical. When the control arms take full responsibility for the suspension loads, all the differential needs to do is transmit the power from the driveshaft to the rear drive axles. For our project Cobra, I chose a late-model Corvette differential. It is lightweight aluminum, but it is strong enough to handle 400 horsepower in a 3500-lb. car. It also has a good limited slip unit, and a reasonable selection of ratios.

Our alternative choice for the Cobra would have been a Ford 9-inch. This unit is strong enough but its use on a car with independent rear suspension requires a mounting box to hold the third member in the car. I felt that designing and building this box was not warranted in light of the availability of the Corvette differential.

I modified the Corvette differential to fit into our Cobra chassis by shortening the wing mounts and fitting adapters for mounting the Oldsmobile Toronado drive axles, which I chose because of their ability to change their length or *plunge*.

Figure 11-6. This shows the drive hub that is splined to accept the Oldsmobile Toronado drive axles. This is the same hub as is used with the front knuckle, except for the splines.

Drive Axles

When the rear suspension knuckle location and travel is determined by the rear control arms, its distance from the differential is always changing. This changing distance requires drive axles that can continuously change their length. This characteristic is called *plunge*. Our need for this plunge feature, along with the need for very strong shafts and joints, required the use of the older-style Oldsmobile Toronado drive axles. These units are available through rebuilders. I shortened ours to get them to fit the narrow Cobra body width (Figures 11-6 and 11-7).

Knuckles

The knuckles used on an independent rear suspension must have provisions for running a drive axle through them. Fortunately, the knuckles I used on our front suspension have this feature, so I was able to use the same parts. These parts are made from cast stainless steel, so they are durable and stiff. The same hubs are used as on the front except they are splined to accept the Toronado drive axles. I was also able to use the same brake rotor diameter as on the front, but I did use smaller piston sizes in the rear calipers. Both the calipers and the rotors bolt right on the knuckle and wheel pilot pieces.

Control Arms

Because of the similarity between the independent front suspension and the independent rear suspension, I was able to use the same basic style control arms, front and rear. At the rear, the upper arm is offset to clear the spring shock unit. The rear lower arm is also different where the spring/shock attaches to it.

The position of the upper arm on the frame is different at the rear because of the need to provide

Figure 11-7. Individual drive axles are needed on a car with independent rear suspensions. If the suspension is the formula type, these shafts must be able to accept changes in length as the wheels move up and down. Oldsmobile Toronado drive axles provide this feature together with very high load capabilities.

anti-squat geometry. At the front, there is anti-dive, which tilts the upper arm in a different direction. In order to have the correct deflection steer characteristics at the rear, I mounted the tie-rods behind the axle. This arrangement produces toe-in during cornering, which makes for more consistent and predictable handling. At the rear, the steering gear is replaced with tie-rods that anchor to the frame. These frame anchor points are positioned to give a slight amount of roll understeer.

Bushings

Control arm bushings have a dramatic effect on the handling precision of a front suspension system. On our independent rear suspension system, they are even more important because of the added forces generated from driving the rear wheels. As explained earlier, slight changes in toe-in or camber have a much greater effect at the rear than they do at the front.

If rubber bushings are selected for noise and harshness considerations, they should then be arranged so the deflection has a minimum effect on the alignment of the rear wheels. All of the bushing considerations used on the front suspension should be applied to the rear suspension. ■

FRAME DESIGN 12

Good handling requires adequate chassis stiffness. This means the car's frame structure, or the unit body, must be rigid enough to resist bending and twisting. Designing a rigid frame with or without a roll cage structure requires the application of a few basic elements. A maze of steel tubing might be visually impressive, but unless the tubes are properly arranged and secured, they will not produce a rigid structure. The elements of a structural design are not difficult to master. It is possible to build models to test your frame and/or roll cage design. By evaluating which shapes are inherently rigid and which are flexible, you can visualize how your car's chassis will deflect and bend under the loads produced by driving conditions.

Remove a race car's bodywork and you'll be confronted with a maze of bars and tubes that comprise the frame. Chassis stiffness is an essential ingredient to good handling. The frame must be designed to withstand the high bending and twisting forces it will be subject to. However, frame design and construction is more than just welding together a bunch of tubes.
Photo by Michael Lutfy.

STRUCTURAL BASICS

Before actually designing a frame and/or roll cage, it is necessary to recognize what shapes and arrangements are rigid and which are not. The basic shape for constructing rigid structures is the triangle. Its shape and dimensions will not change much unless one of its three legs is broken (Figure 12-1). In contrast, a square-shaped set of tubes has very little structural rigidity, in that it will bend diagonally when even a small load is applied to it (Figure 12-2).

Figure 12-3 shows a rectangular shape braced with a diagonal member. This divides the rectangle into two triangles, so that a shape which was very weak becomes rigid. Double diagonals can be used (which creates four triangles) for still more rigidity, but these additional members are usually unnecessary unless very high loads are anticipated.

Figure 12-4 shows the use of a panel of thin metal (cardboard in our model) to give the rectangle diagonal rigidity. This is called a *shear plate* and its

Figure 12-1. The triangle is the basic shape for creating rigid structures. Its dimensions cannot change unless one of its three legs is broken. There are two vital considerations in automotive structural design: Will the structure break, and will it deflect under load?

effect is the same as a diagonal brace. Shear plates can be used to advantage in race cars, because they can function as firewalls, floorboards and bulkheads, thus eliminating the weight and complexity of diagonal tubes.

Figure 12-2. This common arrangement of tubes—either rectangular or square—is very weak diagonally. Only the strength of the joints is available to resist movement.

Applying the rigidity analysis from the two-dimensional examples to a three-dimensional box reveals how the basic structure of a car can be improved. The most difficult forces to resist in a chassis are the loads that put the frame in *torsion*. Twisting an open

CHART 12-1

Wheelbase	101.0 in.
Overall Height	48.0
Overall Width	56.0
Overall Length	60.0
Frame Ground Clearance	4.0
Front Spoiler Ground Clearance	5.00
Tire O.D.	26.00
Engine Location	
Ground to E Crank	10.0
Front Spark Plug to Rear of Block	18.7
Thickness of Engine Plate	.3
Set Back Dimension	2.0
Distance from E Front Axle To Front Frame Bulkhead	21.0
Rocker Frame Rail Size	2 x 3 x .12 wall
Fuel Cell Size	9 x 17 x 33
Height above Ground	11.0

You'll need these dimensions to layout your frame. The numbers here reflect our sample layout covered later in this Chapter. First, you need to know some structural basics.

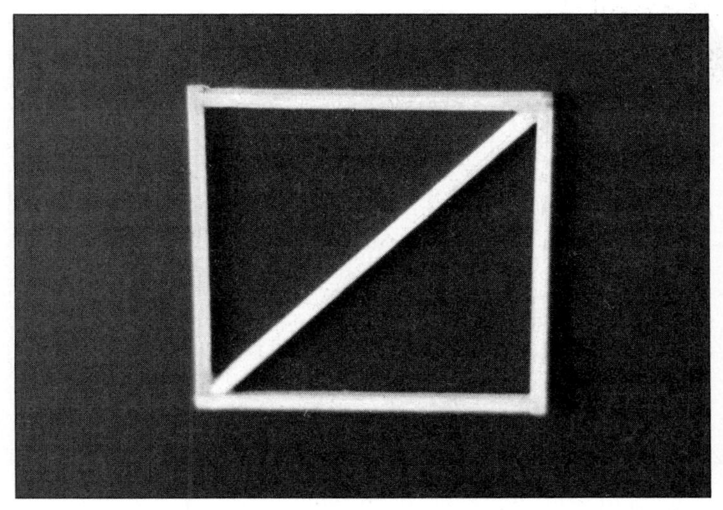

Figure 12-3. Here, the basic rectangular shape is braced with a diagonal member. This divides the rectangle into two triangles-thus a shape which was very weak becomes extremely rigid. Double diagonals can be used (which creates four triangles) for still more rigidity, but these additional members are usually unnecessary unless very high loadings are anticipated.

Figure 12-4. Here a panel of thin metal (well, actually cardboard in our model) is used to give the rectangle diagonal rigidity. This is called a "shear plate," and its effect is the same as a diagonal brace. Shear plates can be used to advantage in race cars, because they can function as firewalls, floorboards and bulkheads, thus eliminating the weight and complexity of diagonal tubes.

box clearly shows how poorly it absorbs torsional loads (Figure 12-5). This is what happens to areas of an automobile that are not triangulated. Even though five of its six sides are rigid, the one open side makes the whole assembly very weak.

Just as a single diagonal brace strengthens a two-dimensional rectangle, a diagonal across the open box makes the total package torsionally rigid (Figure 12-6).

Figure 12-5. Applying our rigidity analysis from the two-dimensional examples above to this three-dimensional box reveals how the basic structure of a car can be improved. The most difficult forces to counter in a chassis are the loads that put the frame in torsion. If these torsional loads are handled, then the bending loads are usually handled as well. Twisting this open box clearly shows how poorly it absorbs torsional loads. This is exactly what happens to areas of an automobile that are not triangulated. Note that even though five of its six sides are rigid, the one open side makes the whole assembly very weak.

Stiffness vs. Strength

Designing a rigid frame is an exercise in applying the structural basics to your application. Each section of the frame must be analyzed to determine how best it can be made rigid. Since most of the loads acting on a car are fed into the front and rear suspensions, frame stiffness between these two points is most important. Notice that I am discussing *stiffness*, not strength. Stiffness refers to how much

Figure 12-6. Just as a single diagonal brace strengthens a two dimensional rectangle, a diagonal across the open box makes the total package torsionally rigid.

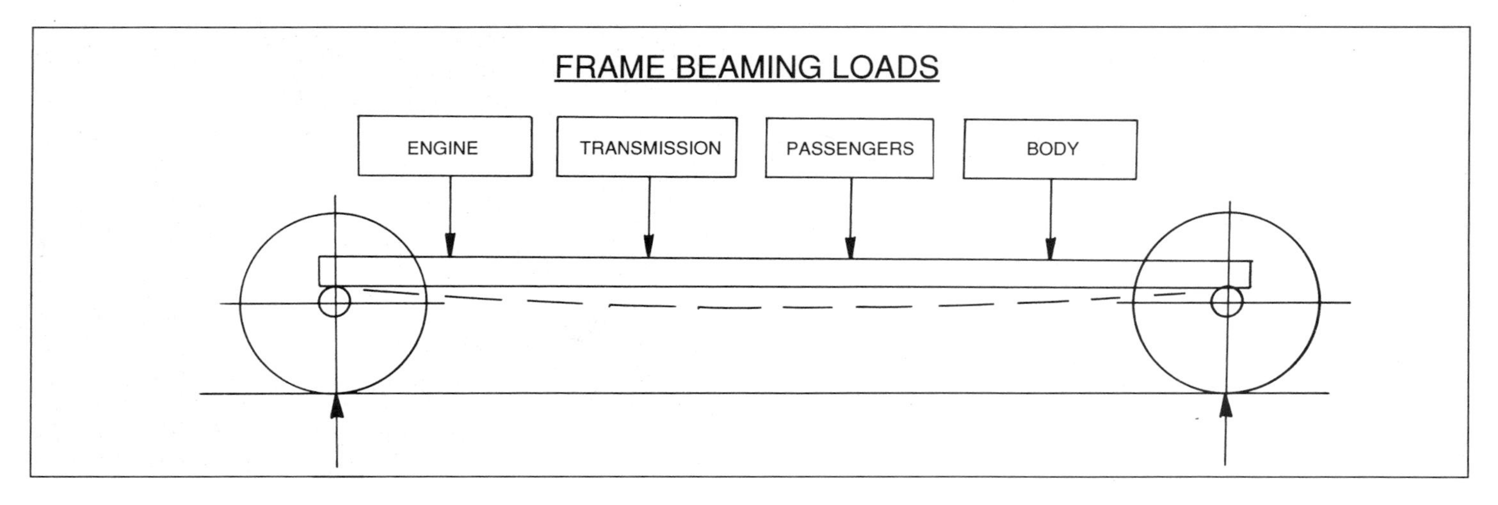

Figure 12-7. Car components can bend the frame because of their weight. It is easy to build a frame that is stiff enough to resist these bending forces in the beaming direction.

something will bend when it is loaded. Strength refers to how much load something can handle before it will break. These are two separate factors, and although they are related, they are not the same. It is possible to have a frame that is strong, but not stiff. For example, your car could run 100,000 miles, and if there were no frame cracks, it would be considered strong enough. But, it would be very possible that this same frame could flex and bend at every turn, which would mean that it wasn't very stiff. There are two aspects of frame stiffness.

Beaming Stiffness—Beaming stiffness refers to how much a frame will flex as it is loaded in the center and supported at both ends. The weight of the engine, transmission, driver and body all cause the frame to be loaded in beaming. Figure 12-7 and 12-8 show a model of a simple ladder frame loaded in beaming (the weight is a 2-oz. nut).

Notice that this amount of weight does not cause much deflection on the model when it is used to load it in the beaming direction. Building a frame that is stiff enough in the beaming direction is not too difficult, so most cars do not have a deflection problem with this type of loading.

Torsional Stiffness—Torsional stiffness refers to how much a frame will flex as it is loaded when one front wheel is up and the other front wheel is down while the rear of the car is held level. This condition is seen at every corner of the road, so its importance to proper handling should be obvious. Figure 12-9 and 12-10 show our same test model loaded in torsion with the 2-oz. weight on a 6-inch long lever arm. The torsional deflection is substantial. This comparison shows how a frame that is stiff in beaming can be very flexible in torsion.

Models—If you don't have the frame yet and you want to determine what design factors can increase its torsional stiffness, models can be very useful. A simple balsa wood and paper model can be a very useful tool in determining what frame configuration gives the best torsional stiffness. My experience has shown that models built to a 1/12 scale using model airplane cement, balsa wood, and heavy paper provide a very useful means of determining the torsional stiffness of a frame. You can build a model of the frame design you want to buy, or build, and then test its stiffness. You can also use the model to determine what modifications can increase its stiffness.

TYPES OF FRAMES

Now let's examine the various types of frame design to determine which may best suit your application.

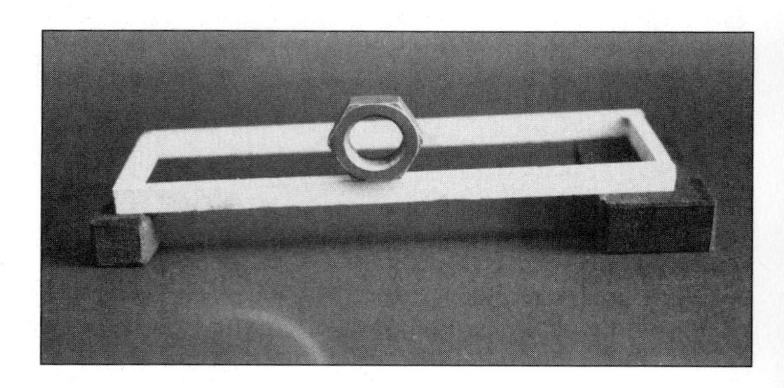

Figure 12-8. Here is a balsa wood model of a simple ladder frame supporting a 2-oz. weight. Note that there is almost no deflection because the frame is stiff enough in beaming to support the weight.

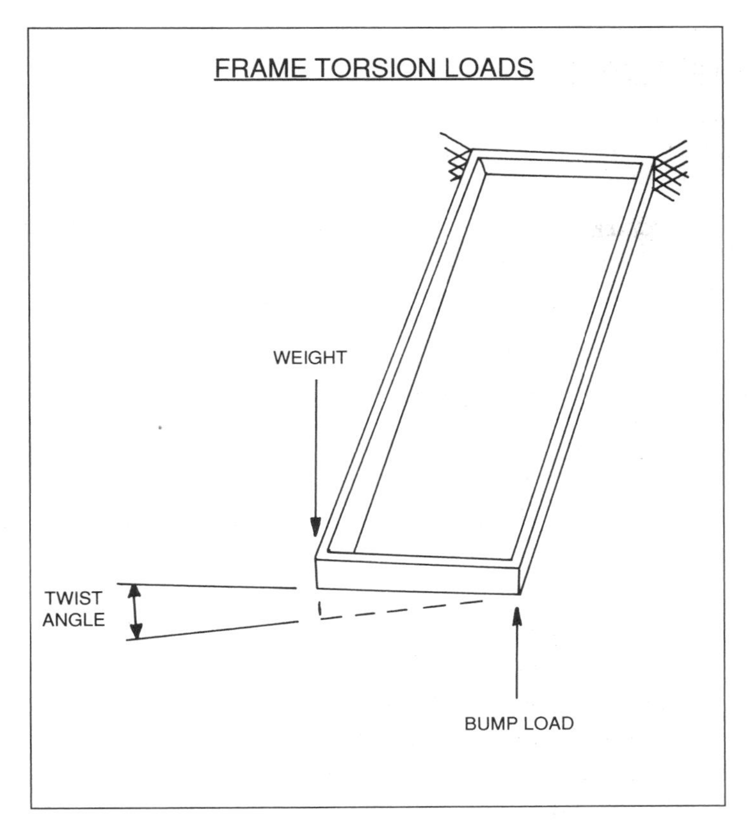

Figure 12-9. Uneven wheel loading over bumps tends to twist the frame. It is difficult to build a frame that is stiff enough to resist these torsional forces.

Ladder Frames

The simplest and most basic type of automobile frame consists of two frame rails connected by two or more crossmembers. This is called a *ladder frame*, because it looks like a ladder. Many cars use a ladder frame, because it is easy to build and it offers good beaming stiffness. Unfortunately, a ladder frame has very poor torsional stiffness, so cars that use this design can have a lot of body creaks

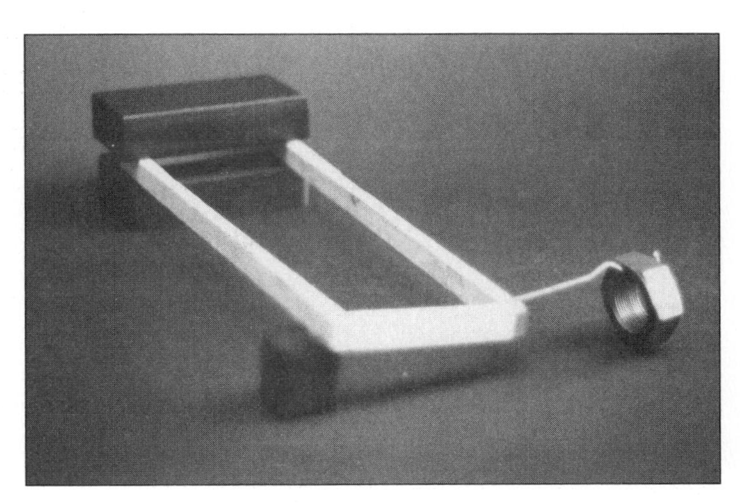

Figure 12-10. This photo shows how much a simple ladder frame will twist when the 2-oz. weight is hung on a 6-in. long arm attached to the front of the frame when the rear is held solid. This twisting force is 12 oz-in. of torque.

and groans as well as a multitude of vibrations caused by the suspension flexing the frame. When a ladder frame is supported by a very stiff steel body shell, the combination can be made fairly stiff in torsion. However, when the body is fiberglass, torsional stiffness is usually lacking. Convertibles with ladder frames are also very poor in torsion, because they do not have the roof structure to help stiffen the total assembly. Cars that use ladder frames include Cobras, Corvettes and many older production cars.

When crossmembers and/or a roll bar are added to a simple ladder frame, there is not much improvement in its torsional stiffness (Figure 12-11).

Some increase in torsional stiffness can be added to a simple ladder frame with the use of an X-mem-

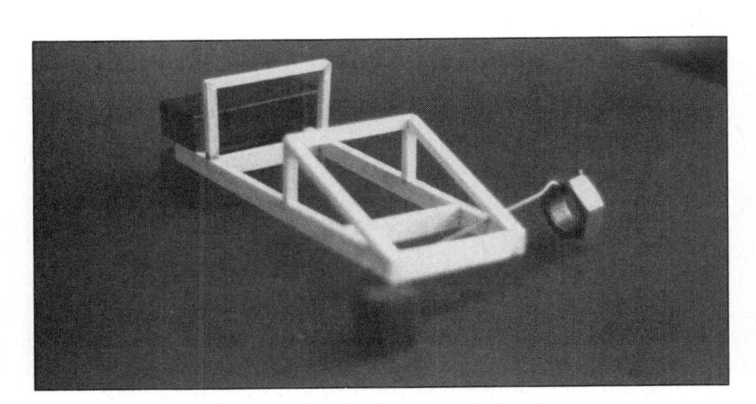

Figure 12-11. This photo shows how much, or how little, torsional stiffness is improved when crossmembers or other structural parts are added. As can be seen, adding crossmembers does little to improve torsional stiffness.

Figure 12-12. This photo shows that adding an X-brace does make some improvement in a ladder frame's torsional stiffness. Corvette convertibles use this design but, as can be seen, there is still some frame twisting.

Figure 12-13. Adding a simple roll cage to a ladder frame does not add much to the overall torsional stiffness unless the whole system is designed to absorb the twisting forces.

Figure 12-14. This photo shows the same simple roll cage on a ladder frame with just a few added tubes strategically placed to improve the torsional stiffness of the total structure. As can be seen, the improvement is dramatic.

ber between frame rails. Figure 12-12 shows how this modification increased the torsional stiffness of our sample model. The late-model Corvette convertible uses this technique to improve frame torsional stiffness.

Ladder Frame w/Roll Cage—Adding a full roll cage to a simple ladder frame can result in a rigid structure. Surprisingly this is not always the case, as is shown by this example. I built a frame model based on the minimum SCCA requirements (Figure 12-13). When this model was tested for torsional stiffness, it was apparent that it did not have much. It should be noted that the SCCA roll cage has evolved over the years, with its primary goal being safety. By adding a few extra tubes in the correct locations, I found it was possible to greatly increase the torsional stiffness of the basic SCCA roll cage on a simple ladder frame (Figure 12-14).

Figure 12-15. This photo shows that even a massive NASCAR roll cage does not necessarily add torsional stiffness to a ladder frame.

Bulkheads—Another way to improve the structure of the roll cage is to make a front and rear bulkhead in front of and behind the driver. In order for the roll-over tubes and the side-bar tubes to be effective, they have to be mounted on rigid members. The front and rear bulkheads offer this solid foundation. They can be made from square and rectangular tubing and then braced and triangulated with steel shear panels that are welded to the tubes. The front bulkhead can also be supported with an engine mounting plate, which bolts into the large hole where the engine mounts. The rear bulkhead can be supported with a bottom tube, which also acts to retain the driveshaft should it come loose. This type of bulkhead construction offers improved lateral strength to the chassis while providing an excellent base onto which roll cage tubes can be mounted. After the roll cage tubes have been mounted on the front and rear bulkheads, front and rear frame snouts can be added to support the suspension

Figure 12-16. When shear panels and a few extra tubes are added to the regular NASCAR roll cage, there is a dramatic improvement in the torsional stiffness of the structure.

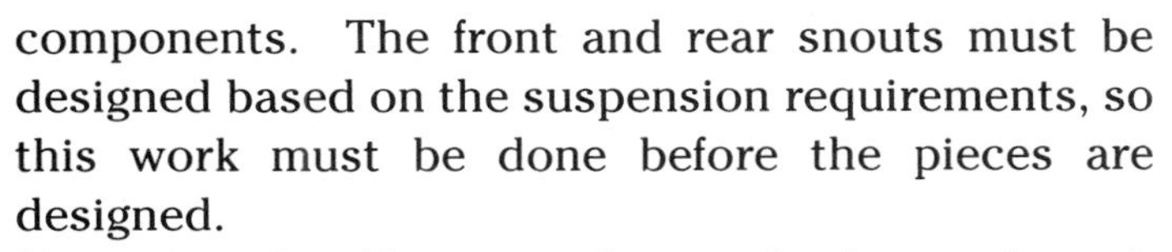

Figure 12-17. A backbone frame such as the one shown in this photo is a structure that is very stiff in torsion. Since it is also a simple design, it is lightweight and can be used for street-driven two- and four-passenger cars.

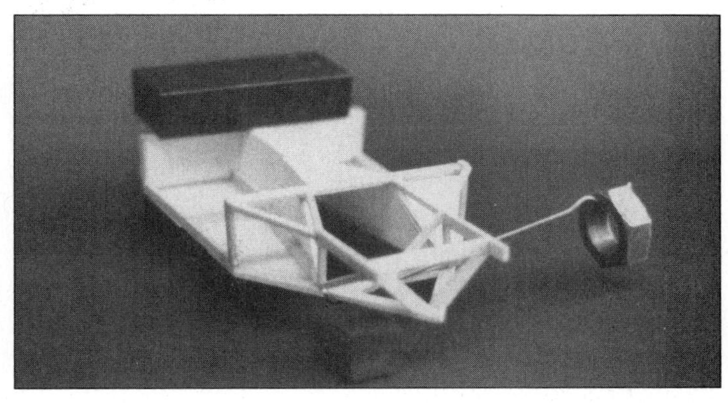

Figure 12-18. Here is a frame model which has a smaller backbone to provide more passenger space inside the car. With appropriate front and rear subframes, this design provides a simple and lightweight frame structure that has excellent torsional stiffness.

components. The front and rear snouts must be designed based on the suspension requirements, so this work must be done before the pieces are designed.

Shear Panels—Shear panels can also be used to triangulate the floorpan and tunnel surfaces. Steel panels must be used in this location anyway, so why not get some structural benefit from them? I built a frame model to show how the bulkheads would combine with a standard NASCAR roll cage (Figure 12-15). Structurally designed floorpan and tunnel panels were then added to this model to test their effect on torsional stiffness. Figure 12-16 shows the model with all these shear panels as well as the tubes we added for better torsional stiffness. As can be seen from the photo, these changes have resulted in a chassis that has much greater torsional stiffness.

Backbone Frame

The original Volkswagens, many Lotus's and the new Jackrabbit Kit Car use a frame design that provides a solution to the problem of torsional stiffness. This frame design is practical for two-passenger and four-passenger cars, because the center backbone can be used as an armrest or console between the seats. Figure 12-17 shows the testing of a model that was used to build a Trans-Am race car. This car has

Figure 12-19. This photo shows a full-size backbone frame that is designed to fit inside a standard Cobra body shell. It is made from aluminum and weighs less than 200 lbs.

a very large tunnel and is extremely stiff in torsion.

A more practical backbone frame design is shown being tested in Figure 12-18. The size of this backbone is smaller, but because of its configuration, it is still very stiff. Notice in the photos that there is some deflection (about 1 degree). Since this deflection is about 1/6 of what was measured with a simple ladder frame, a car using this design would have 6 times the stiffness. A backbone frame can be constructed from flat stock, from a large piece of tubing, or from a collection of tubes if they are properly arranged.

The model shown in Figure 12-18 was eventually built into a full-size prototype. Figure 12-19 shows this frame as it was constructed from aluminum sheet stock and aluminum tubing. This total frame structure, including the floorpan, the tunnel, the door posts and the suspension pickups, weighs less than 250 lbs. Figure 12-20 shows this same backbone frame with the suspension and driveline attached.

Not every car can use a backbone frame because of the space or seating requirements. Every car, however, can have its frame stiffness evaluated using the model technique. If you have any concerns about the stiffness of your existing or future frame design, build a model of it and test it.

Seating Location—The way to increase the torsional stiffness of a ladder frame is to make the frame rails and the cross-member stiffer. This has limited effect, because when the members become bigger, they get heavier and there is less room for people. If a simple ladder frame is used, the driver and passenger sit on top of the frame like on a normal Cobra (Figure 12-21). The late-model Corvettes solved this problem by running the side rails outside of the passengers so the seating position can be lowered (Figure 12-22). This modification resolves the seating position problem but it does nothing to solve the torsional stiffness problem. This can be seen when a model of the late-model Corvette frame is tested, even with an X-member like the one used on the Corvette convertible.

FRAME LAYOUT

A good way to visualize your car's frame and roll cage requirements is to make a scale drawing of the entire car. I used a 1/12 scale to keep the paper size within reason. This overall car layout should show the body outline, the wheel size and location, the driver position, the engine location and the fuel cell location. These components are not likely to change, so you must design the frame and roll cage to fit around them.

I used a side-view photo of the car body to get the body outline and the seating position. Next, I measured the actual tires and engine to determine their size for the layout. The specifications, in inches, used to make the overall car layout (Figure 12-23) are shown in Chart 12-1, page 86.

Since each section of the frame and roll cage structure must be fully triangulated to achieve a rigid structure, it is important to have as few intersection

Figure 12-20. Here is the same aluminum backbone frame with the suspension added. Notice how the loads from the front suspension follow the tunnel to the rear suspension. This load path is what provides the torsional stiffness.

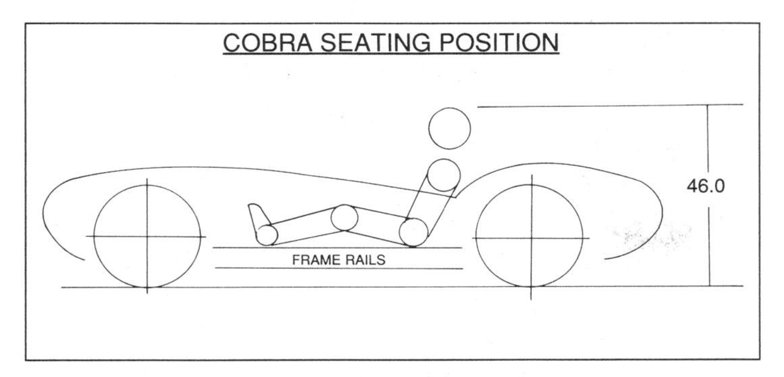

Figure 12-21. Large frame rails and an X-brace require the driver and passenger to sit on top of the frame. This obviously raises the seating position, so they look like they are sitting on top of the car, not inside it.

LOWER SEATING POSITION

42.0

Figure 12-22. If the rails are bowed out to go around the driver and passenger, the seating position can be lowered. The newer Corvettes use this design but since the X-brace must be below the frame rails, there is reduced ground clearance.

points as is practical. After reviewing the overall car layout, you can determine that there will need to be 3 separate sections to the structure. These will be from the: front suspension cross-member to the front cowl bulkhead; front cowl bulkhead to the main roll bar loop behind the driver; and the main roll bar loop to the rear cross-member.

By selecting the ideal locations for the front cowl bulkhead and the main roll bar loop, we have simplified the frame. If a location was chosen for the front cowl bulkhead that was different from where the front of the roll cage side tubes are located, extra triangulation in the structure would have been required to tie these two intersections together. ■

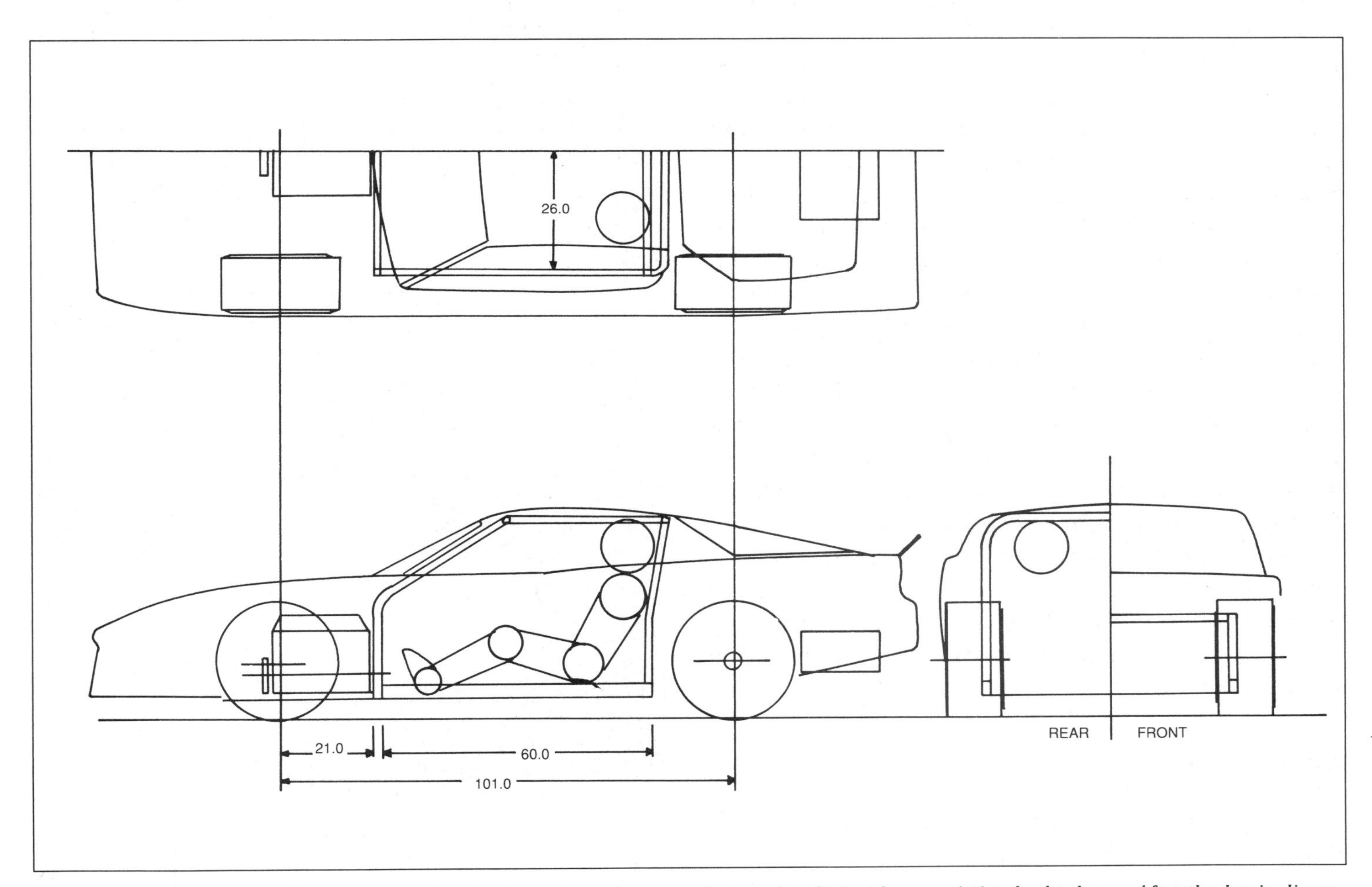

Figure 12-23. This outline drawing shows how a frame layout can be designed to fit inside an existing body shape. After the basic dimensions are established, you can build a model of the frame structure to determine its stiffness.

BUILDING A FRAME 13

A frame should not be constructed until all of the design work is completed. You should have your layout completed and your test model fully developed before any metal is cut. It is normal to make dimensional changes as you build the frame, but these changes should be made within the design considerations you have already established. The following example reviews some of the design and fabrication steps taken in building a short-track oval race car frame.

FRONT & REAR BULKHEADS

One of the design features previously discussed, was the use of front and rear bulkheads ahead of and behind the driver to add structure in these areas. A roll cage and side bars are only as strong as the members they are mounted on, so building strong bulkheads is recommended. Rigid bulkheads also add to the frame stiffness, so there is a double benefit from their use.

After you carefully develop and test your frame design, it's time to put all the pieces together. Ideally, you should use a welding jig to keep everything square and level. However, a level garage floor can suffice as long as you are very careful to keep everything aligned.

Figure 13-1. Making the front and rear bulkheads from square or rectangular tubing results in a very stiff member. This photo shows an adjustable control arm bracket installed prior to the shear panel.

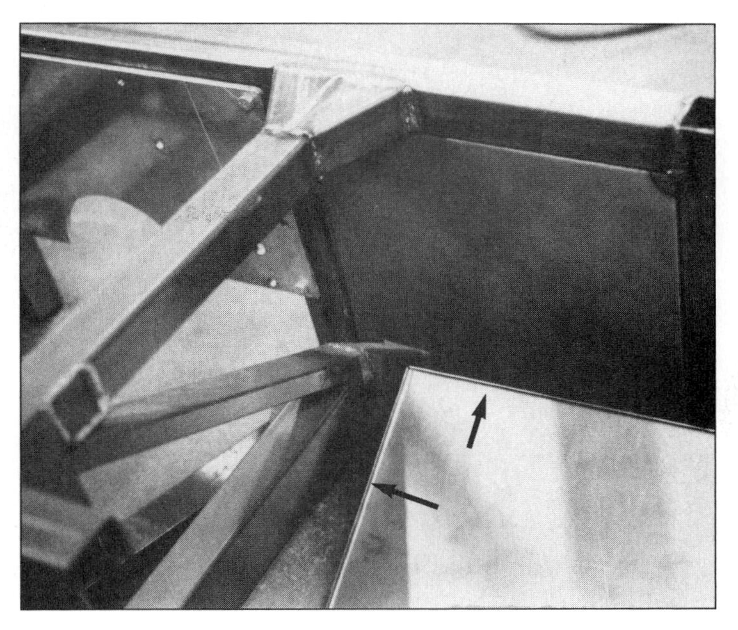

Figure 13-2. A small flange(arrows) can be formed on the panels to provide an easy weld edge where they meet the tubes. Note how the engine mounting plate at left is used to stiffen the center opening.

We made the bulkheads for our project car from 2.00 x .095 in. square tubing. Square tubing is much stronger and stiffer in bending than round tubing, and its use permits a convenient means of attaching shear panels, reinforcing plates and brackets.

In order for a shear panel to be effective, it should be welded all the way around. To prevent warping at the weld, we put a 1/4-inch flange at each edge. This hard corner keeps the panel flat during welding and results in a part that is not only rigid, but also good-looking.

To attach a reinforcing plate, such as an engine plate, use *Rivnuts*. These treaded inserts make assembly and attachments easy and convenient. Bolts stay tight in them, and they can be replaced if need be. The use of an engine plate bolted to the front bulkhead serves two functions: one is to mount the rear of the engine, the other is to provide full triangulation of the front bulkhead. The shear panels triangulate the outboard ends of the front bulkhead, but the engine plate is needed to stiffen the center position (Figure 13-1 & 13-2).

FRONT & REAR SNOUTS

After the front bulkhead is completed, the front snout can be added. The front snout must rigidly hold all the front suspension and steering components, so it must be designed to resist the very large forces involved. The forces on the front control arms can exceed 5000 lbs., so it is important to design the mounting points to be as strong and rigid as possible. Adding a lot of heavy metal is not the answer; arranging the brackets and tubes to provide a rigid structure is. With straddle-mounting brackets and triangular arrangement of the tubes, it is possible to build rigid front and rear snouts that are not any heavier than conventional ones. Building a model of the front and rear snouts is a good way to

Figure 13-3. This photo shows a completed front snout with all the suspension and steering brackets installed. Note that the front bulkhead has been incorporated into this structure.

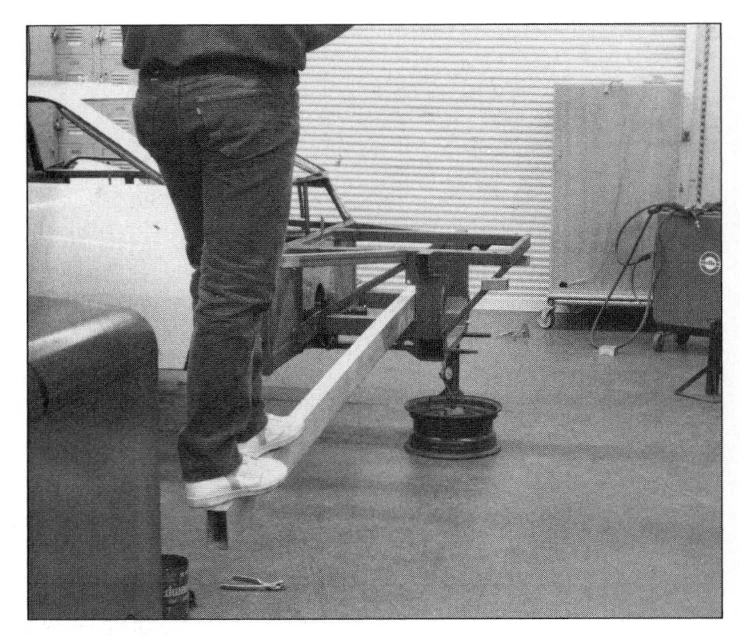

Figures 13-4, 13-5. The actual torsional stiffness of a completed frame structure can be measured by twisting the frame with a known torque. This setup uses a dial indicator to determine the twist with a 2000 lbs-ft. torque applied to the front of the frame.

determine if they are rigid, and to see if the central frame structure is arranged to properly distribute the torsional loads from the front suspension pickups to the rear suspension pickups. Remember, if something can bend, it will, and when it bends some part of the suspension or steering is going to move even more. When this happens, the car is going to do something the driver doesn't want or expect it to, so he won't be able to drive it as fast as it can go (Figure 13-3).

CENTRAL FRAME STRUCTURE

The central frame structure must connect the front and rear suspension snouts in a rigid manner while providing the driver with maximum crash protection. It is also important to build this central structure so that the front and rear suspension are in alignment with each other. Ideally, this is accomplished on a frame fixture or *jig* which has been leveled and squared. However, a level and square frame can be built on your garage floor if you are careful to keep everything in alignment. Measure everything twice. You can use shims to level the frame members on the floor and cross-measure to keep things square. It takes a lot longer than using a frame fixture, but the results can be just as good. Like so many things, how you use your equipment is more important than what equipment you use. One trick you can use to minimize the warping that comes from welding is to tack-weld all four corners of all the tubes and then recheck the alignment. When this is done, the final welding usually doesn't cause too much distortion.

We used our frame model to determine the exact arrangement of the diagonal tubes and shear panels to satisfy the rules and to provide adequate room for the driver and the driveline. There is no right way or wrong way to do this. The important thing is to end up with a rigid structure that meets your requirements.

MEASURING TORSIONAL STIFFNESS

If the frame and roll cage are built to the same configuration as the model with good torsional stiffness, the resulting structure should also be torsionally stiff. We tested our project car frame and roll cage to measure its torsional stiffness using a setup similar to the one we used to test the model. The rear of the frame was anchored so it couldn't move, and then a twisting force was applied to the front of the frame. The twisting forces were applied by our 200-lb. driver at the end of a 10-ft. beam. This resulted in a torque input of 2000 lbs-ft. We supported the left front corner on a jackstand, and used a dial indicator to read the deflection on the right front. The total deflection was .068 inches as measured 20 inches from the support. To find the frame torsional stiffness, we used this formula:

$$\text{Torsional Stiffness} = \frac{\text{Torque}}{57\text{ deg.}} \times \frac{\text{Spread Distance}}{\text{Deflection}}$$

$$\text{Torsional Stiffness} = \frac{2000\text{ lbs-ft.}}{57\text{ deg.}} \times \frac{20\text{ in.}}{.068\text{ in.}}$$

Torsional Stiffness = 10, 319 lbs-ft./degree

The 57 degree value is needed to convert the vertical deflection into an angular measurement (Figure 13-4 & 13-5). Over 10,000 lbs-ft./degree of torsional stiffness is very good, so our model helped us build a stiff frame and roll cage. The total weight of our frame and roll cage with all the brackets and shear panels is 392 lbs. Most of this weight is at the rear and on the left side where any reduction in weight would have to be replaced with ballast, so there would be no performance disadvantage as long as the total weight stays under the limit. We can reduce the rear weight by about 50 lbs. once we complete our rear suspension tests and remove some of the brackets and frame braces.

Crash Protection

Crash protection for the driver is one of the frame and roll cage's primary purposes. By making the total structure stiffer, it becomes stronger in a crash situation. Our bulkhead design improves driver protection front and rear just as the roll-over tubes and side bar tubes do on the top and at the sides.

Top fuel dragsters also have used roll cages successfully to protect the driver. The roll cage they have developed over the years includes a roll bar loop around the drivers head as well as behind it. This design permits the use of panels and insulation that can support the sides of the driver's head just as the head-rest does at the rear. When a driver's head is not supported at the sides, his neck can be injured if the car is T-boned. ■

Top Fuel dragsters have used roll cages successfully to protect the driver. A roll bar is located at the rear and sides of the driver's head. Photo by Michael Lutfy.

AERODYNAMIC DOWNFORCE

14

Aerodynamic downforce is developed on a car when the air pressure on the top of the car is greater than the air pressure on the bottom of the car. If the air pressure on the top is less than on the bottom, the car will generate *lift* just like an airplane. Once this pressure differential is established, its effects will always be in the downforce direction. The amount of downforce (measured in lbs.) will, however, change as the speed changes. The amount of downforce (or lift) a car has will change in magnitude in proportion to the speed squared. An example of this relationship shows that the downforce will be four times as great when the speed is doubled (Chart 14-1).

Chart 14-1

Speed	Downforce
50 mph	100 lbs.
100 mph	400 lbs.
200 mph	1600 lbs.

AMOUNT OF DOWNFORCE

Downforce can be increased on any car body by causing a higher pressure on the top of the body and causing a lower pressure on the bottom of the body. To determine the amount of downforce, you'd simply multiply the length by the width to get the plan view area in square inches. If a pressure differential of 1/10 pounds per square inch (psi) was caused between the top and the bottom of the car that had an overall length of 192 inches and an overall width of 72 inches, the total downforce would be:

Downforce = Length x Width x Pressure
Downforce = 13, 820 sq. in. (Plan View Area) x .10
Downforce at 1/10 psi = 1,382 lbs.

While Indy-type cars can achieve downforce values this high or higher, stock-bodied cars are typically less than this. For analysis, I will use a total downforce value of 400 lbs.

Managing the flow of air into, over, around and underneath the car is crucial for top-speed handling and performance. Photo by Michael Lutfy.

Chart 14-2

Tire Location	Static Weight on each tire	Lateral Weight Transfer	Vertical Load (weight on tire during cornering)	Cornering Force (from Tire Performance Curve, p. 2)
Left Front	750	-500	250	450
Right Front	750	+500	1250	1130
Left Rear	750	-500	250	450
Right Rear	750	+500	1250	1130
Totals	3000		3000	3160

Cornering Force = 3160/3000 = 1.05 g's

How Downforce Increases Cornering Power

Assume a 3000-lb. car with 50% of the weight on the front and 50% of the weight on the left side. Also assume that the lateral weight transfer due to cornering is 1000 lbs. This weight transfer will be split so 50% is on the front end. Once again, I will use the standard tire performance curve from page 2 to determine cornering power. Under these conditions, the tire loadings and cornering force could be determined as was done in Chapter 2. The total cornering force, 1.05 *g*'s, was determined as shown in Chart 14-2.

Now, if you take the same example and add 400 lbs. of downforce (100 lbs. at each wheel), you'll have results as seen in Chart 14-3, which says that adding 400 lbs. of downforce can increase the cornering power by 10 percent to 1.15 *g*'s.

How Downforce Affects Lap Times

In order to simplify this calculation, let's assume we have a quarter-mile track with no bank angle and a constant 210-ft. radius. The time (T) to run this track at a constant 1.05 *g* cornering force could be determined by the following formula:

$$T^2 = \frac{1.225 \times 210}{1.05\ g}$$

$T^2 = 245$

T = 15.6 seconds

Repeating this calculation for 1.15 *g*'s of cornering force shows that the lap time decreases to 14.9 seconds. We recognize that tracks aren't flat nor do they have a constant radius. But, if the car can run the corners faster, it will be able to run the straights faster, so the effect on lap times will be similar.

INCREASING DOWNFORCE

It should be noted that many cars do not have aerodynamic downforce, they have lift. Older model production cars had considerable lift, especially at the front. It was typical to see 300-lbs. of lift on the front of cars like 1970 Camaros. This was what made them feel "light" at 100 mph. Proper control of the airflow around a car can produce downforce, but improper control can cause lift, so be careful to take these factors into consideration when installing any device that can have an aerodynamic effect.

If you accept that causing more pressure on the top of the car will cause more downforce and improve lap times, the next question is how do you get more pressure on the top of the car and less pressure on the bottom of the car? The following are some of the more common ways to accomplish this. Note that all of the following examples are based on a car using stock body panels.

Chart 14-3

Tire Location	Static Weight on each tire	Lateral Weight Transfer	Downforce from Aerodynamics	Vertical Load (weight on tire during cornering)	Cornering Force (from Tire Performance Curve, p. 2)
Left Front	750	-500	+100	350	550
Right Front	750	+500	+100	1350	1180
Left Rear	750	-500	+100	350	550
Right Rear	750	+500	+100	1350	1180
Totals	3000		400	3400	3460

Cornering Force = 3460/3000 = 1.15 g's

Body Rake

When the front of a car is lower than the rear, less air goes under the car. When there is a lack of air to fill the void under the car, it causes a low-pressure area. This low-pressure area under the car increases downforce. Some of the ways to optimize the low pressure under the car are:

1. Make the bottom of the car smooth and flat to keep the airflow smooth.

2. Slightly recess the floor inside the frame rails to build a fence between the low-pressure area and the airflow along the sides of the car.

3. Set the ride height so the rocker panel is 1.00 inch closer to the ground at the front than it is at the back (Figure 14-1).

Front & Rear Spoilers

A front or rear spoiler is really an air dam that causes a higher pressure in front of it and a lower pressure behind it. When the high pressure ahead of

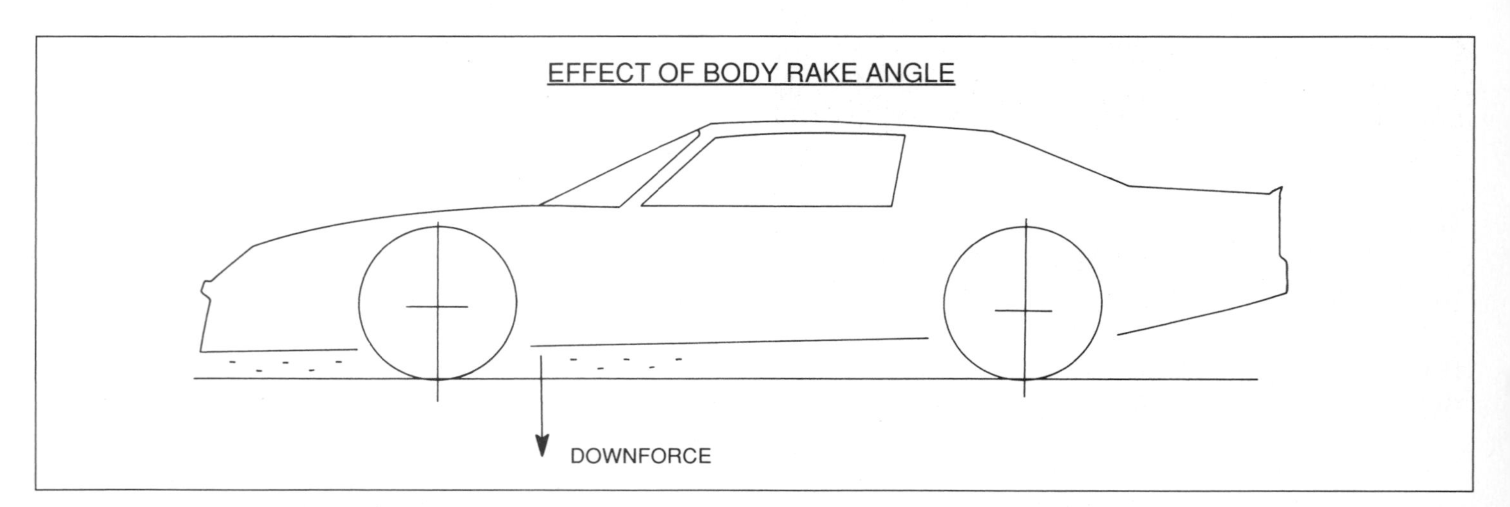

Figure 14-1. If the front of the car is lower to the ground than the rear, a low-pressure area will develop under the car. This low-pressure area will reduce lift and may cause aerodynamic downforce.

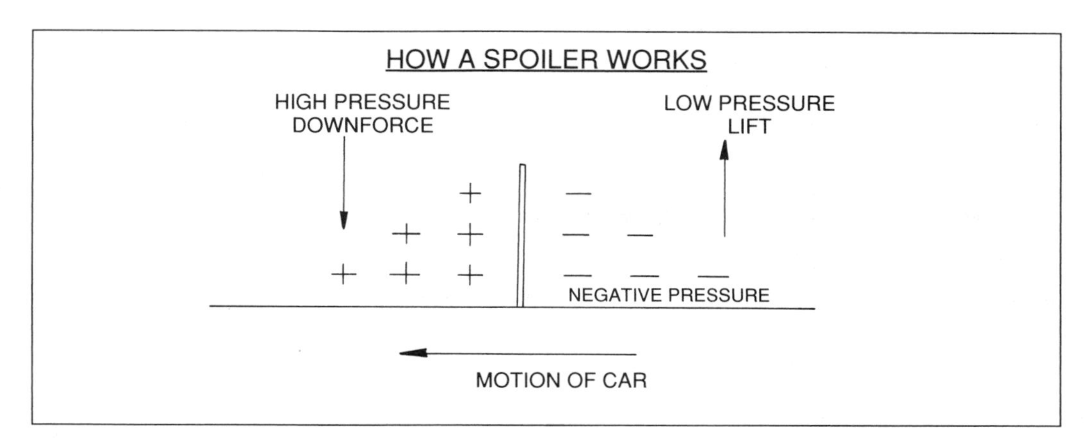

Figure 14-2. A spoiler works by increasing the pressure ahead of it and decreasing the pressure behind it. This positive or negative pressure acts on the surface ahead of or behind the spoiler to produce downforce or lift.

a spoiler can be directed to a horizontal surface on top of the car body, the downforce factor will increase. And, when the low pressure behind a spoiler can act on a horizontal surface on the bottom of the body, it will also increase the downforce factor (Figure 14-2). This is why a front spoiler is mounted at the bottom of the car and a rear spoiler is mounted at the top of the car. These locations maximize the low-pressure area under the car and the high-pressure area on top of the car. The maximum pressure differential on the car is achieved when the front spoiler is mounted vertically at the most forward and lowest possible point (Figures 14-3 & 14-4). The rear spoiler will produce the maximum downforce when it is mounted vertically at the most rearward and highest possible point (Figures 14-5 & 14-6). Obviously, the bigger a spoiler is, the more effect it will have on the pressure differential between the top and the bottom of the car. Bigger spoilers cause more drag, but the increase in downforce they produce is usually more important to total car performance.

Wings

On an airplane, the wing is used to create lift. On a race car, the wing is used upside-down to create downforce. The basic function of a wing on a race car is to create positive pressure on the top surface and negative pressure on the bottom surface. When a wing is attached to the body of a car, the downforce generated by the wing, or wings, adds to the downforce generated by the body panels. Many of the same tricks that are used to cause high pressure on the top surfaces of the car body are used to create high pressure on the top surfaces of a wing. On slow-speed courses, where downforce is more important than drag, the wings we see in use are more a flat panel with a spoiler at the rear than they are true wing sections. The end plates on a wing are used to keep the high pressure on top of the wing from spilling over the ends and diluting the low pressure on the bottom of the wing. Just as the placement of spoilers is very critical, so is the placement of wings. Obviously, a wing can be

Figure 14-3. Although large front spoilers create drag, the advantage of the increased downforce is more important to overall performance. Photo by Michael Lutfy.

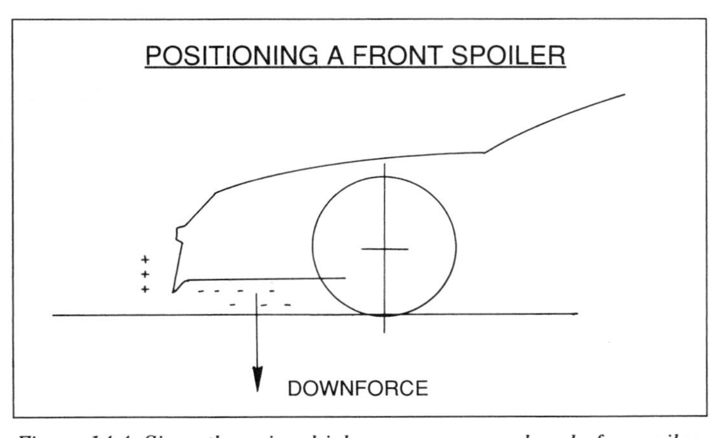

Figure 14-4. Since there is a high-pressure area ahead of a spoiler, and a low-pressure area behind it, a front spoiler should be mounted as low and as far forward as possible. This location keeps the high pressure from acting on any horizontal surface under the front of the car while allowing the low pressure to act all along the horizontal surface under the car.

Figure 14-5. Note the critical positioning of this rear spoiler. If a rear spoiler is mounted as far back and as high as possible, the high pressure ahead of it will act downward on the horizontal surface of the deck lid. Photo by Michael Lutfy.

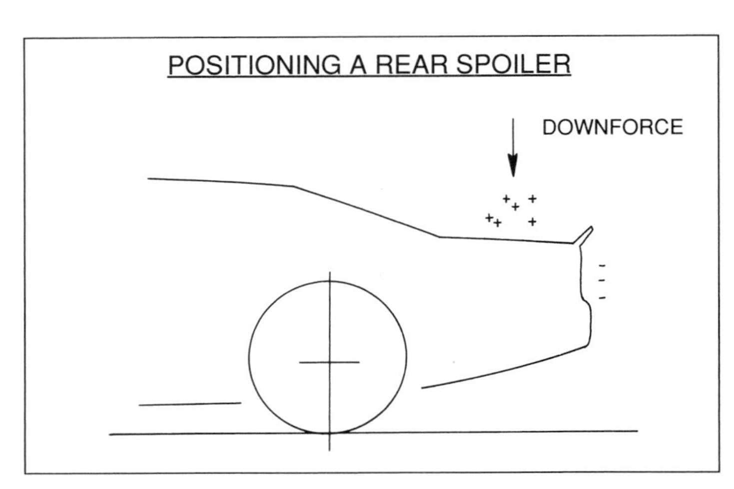

Figure 14-6. If there is nothing behind the rear spoiler, the low pressure behind it will not have a horizontal surface to act upon. Placing a rear spoiler at this location adds to the rear downforce because it only adds positive pressure to the top of the car.

more efficient when it operates in clean air, so it should be mounted where this condition exists. Sometimes we see wings mounted directly above other body panels. This is not a preferred location, because any low-pressure area that is acting against the bottom of the wing is also acting against the top of the body panel adjacent to it. The net effect is obviously diminished since the two forces tend to cancel each other out (Figure 14-7).

Radiator Air Inlet & Exhaust

About 20% of the airflow affecting a race car goes through the radiator and the oil cooler, so how this air is managed is important to both the drag factor and the downforce factor. From a drag standpoint, you want to let the minimum amount of airflow though the grille opening and to the coolers. This means that it is important to be sure that any air that does go through the grille opening also goes through the radiator or the oil cooler. The best way to accomplish this is by building an airtight duct to guide the air between the grille opening and the coolers. Since the grille opening is smaller than the radiator, it is better to mount the radiator at an angle

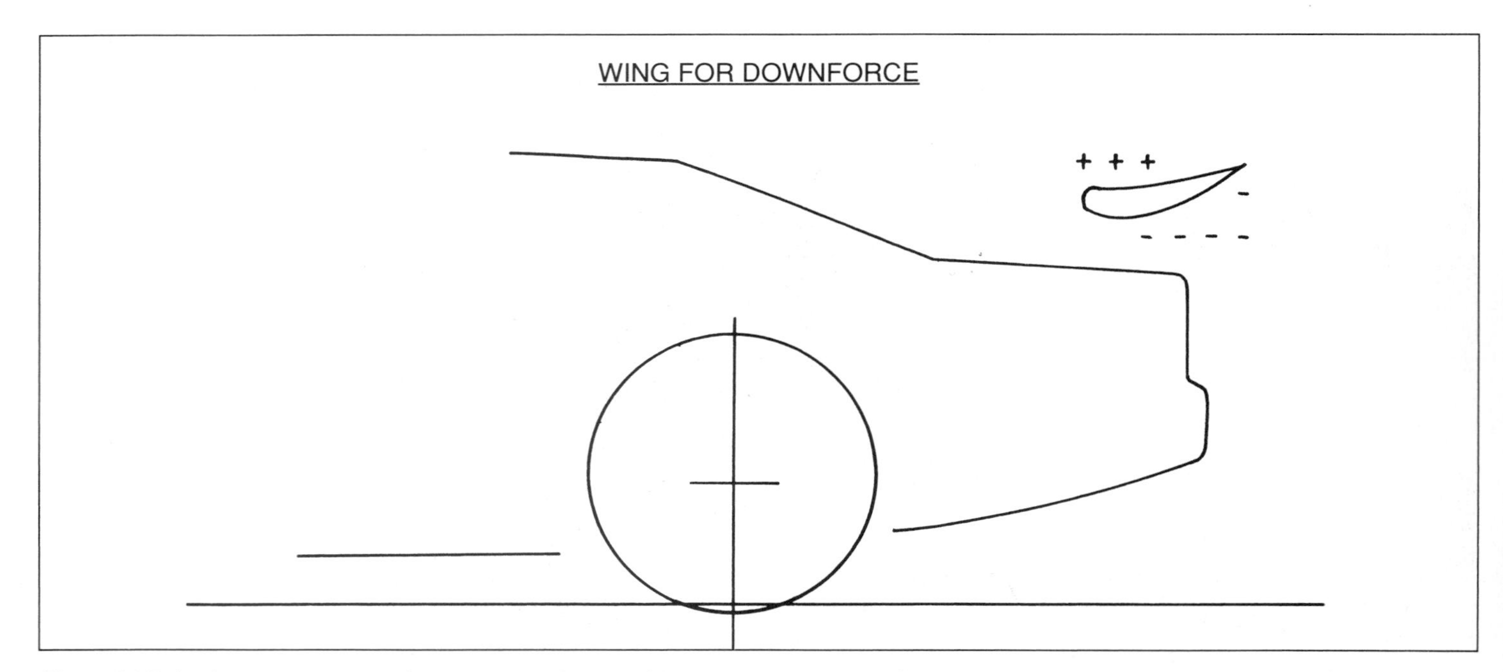

Figure 14-7. A wing on a race car also acts to produce positive pressure on top and negative pressure on the bottom. Many factors such as the size of the wing, its shape, and the angle of attack affect how great this pressure differential will be at any given speed.

The end plates on this rear wing are there to keep the high pressure on top of the wing from spilling over the ends and diluting the low pressure underneath. Photo by Michael Lutfy.

so the air doesn't have to bend up to hit the top of the radiator.

Where the radiator inlet is positioned can have a big effect on front-end lift or downforce. If the radiator air inlet is above the front spoiler, it will see positive pressure and there will be nothing to disturb the low-pressure area behind the front spoiler. If however, the radiator air inlet is behind the front spoiler, two problems exist. One is that the radiator inlet is in a low-pressure area, and the second is that the radiator air inlet disturbs the air behind the front spoiler. Cars that have this arrangement usually require a second spoiler behind the radiator to cause a high pressure area to feed the radiator. Since this spoiler is also behind the front of the car, it causes a high-pressure area under the car which can reduce the downforce or even cause lift.

Even the radiator air outlet or exhaust is important, because dumping this air into the low-pressure area under the car can reduce the downforce. It is better to exhaust the radiator air out the side of the car through the wheelwells, since this will have no effect on the pressure levels below or on top of the car. Ducting can be used to guide the radiator air exhaust to the wheelwells. Keeping the radiator air exhaust away from the engine can help reduce the air temperature at the carburetor inlet (Figure 14-8).

Carburetor Air Inlet

The engine will produce the most power when it breathes the coolest, highest pressure air. At speeds below 100 mph, the cool air is more important than the high-pressure air. If the radiator air ducting is effective in keeping the heated air away from the carburetor, more power will result. NASCAR prohibits using carburetor ducting, but it is possible to feed cool, high-pressure air from the grille opening to the top of the engine so the carburetor can draw it in.

An alternative to this design is to feed air from the base of the windshield to the carburetor. At the center of the car, there is a small high-pressure area where the hood meets the windshield, called the cowl. However, with a sloped windshield, this area is smaller and at a lower pressure, so it is less effective than bringing the air in from the front of the car (Figure 14-9).

Top Fuel dragsters need very efficient wings to keep the front end from lifting during 300 mph runs. Photo by Michael Lutfy.

Note the radiator duct built on the nose of this Camaro Trans-Am racer. Where the radiator inlet is positioned has a big effect on the front-end lift or downforce. Photo by Michael Lutfy.

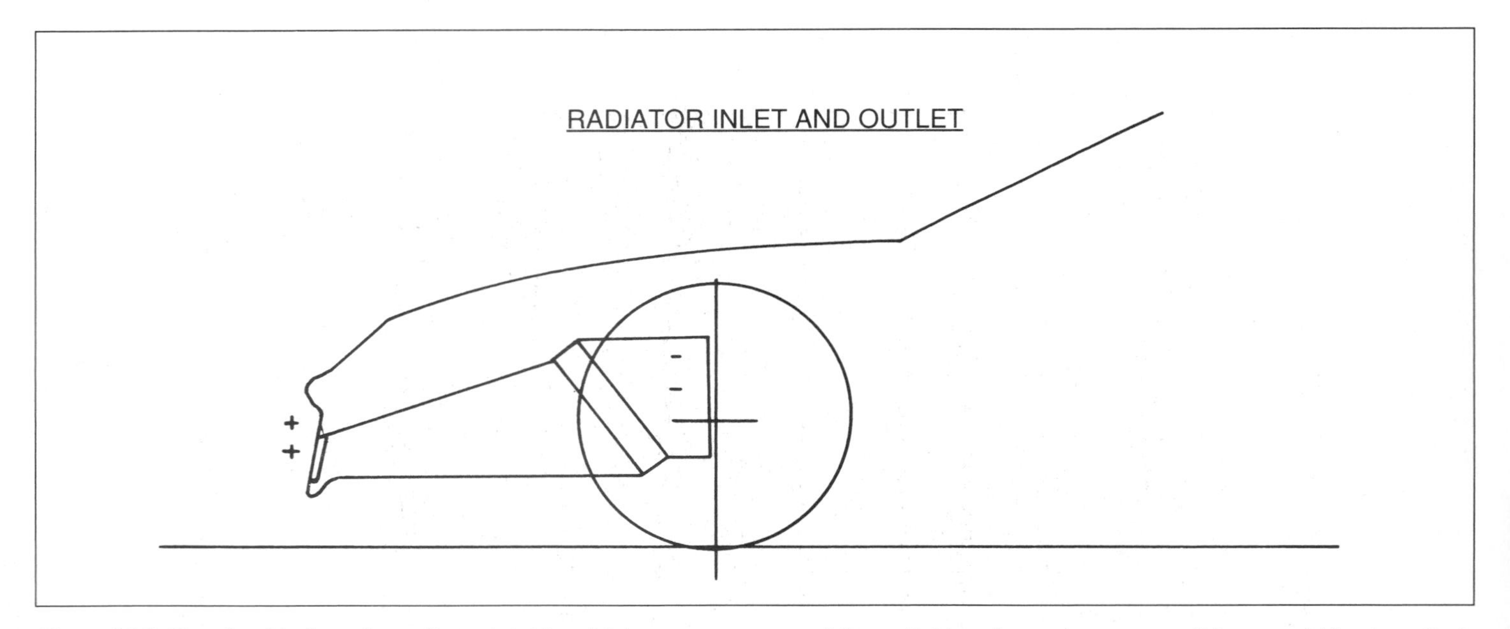

Figure 14-8. You should place the radiator inlet in a high-pressure area and the outlet in a low-pressure area. It is a good idea to exit air out the front wheelwells, because this does not upset any low-pressure area that might be acting on the bottom of the car.

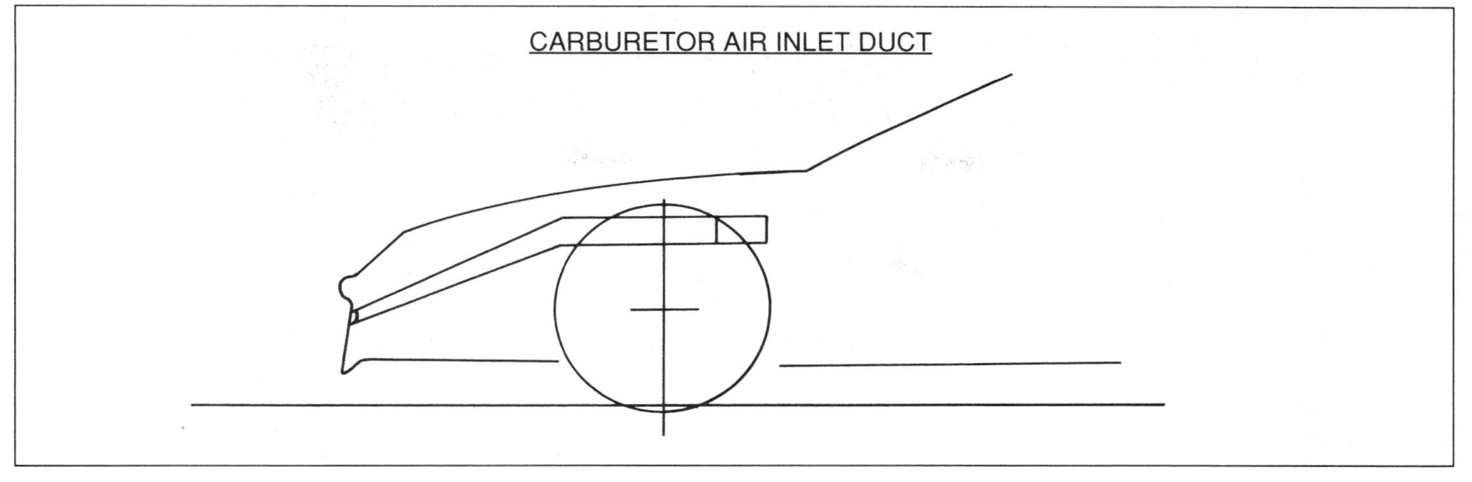

Figure 14-9. Since the engine will develop its maximum horsepower when it breathes high-pressure, cool air, it is best to plumb the carburetor air inlet ahead of the radiator. This location provides the coolest, highest pressure air available.

Balancing Downforce for Handling

If a car has a front and a rear spoiler, or a front and rear wing, it is possible to use the front-to-rear downforce balance to alter the way a car handles. At very high speeds, the downforce balance can have a greater effect than chassis-balancing techniques. The basic procedure is to use more-or-less downforce when more-or-less tire loading and cornering force is needed. Adding front-end downforce will increase the front tire loading and give the front tires more cornering force. When there is no more front-end downforce available, it is sometimes necessary to reduce the rear-end downforce in order to get the car balanced.

There are as many aerodynamic tricks and techniques as there are chassis ones. Each car must be tuned and developed to realize its optimum performance. You don't need a wind tunnel to correctly locate your spoilers, but of course, one would be helpful to optimize the airflow around the entire car. Use some good judgment, and try to analyze where the high-pressure and low-pressure areas will be. If you do this, you'll make progress toward increasing your car's downforce. ■

At very high speeds, such as those generated by Indycars, the downforce balance can have a greater effect on handling than chassis techniques. This is especially true at tracks like Indianapolis and Michigan. Drivers and crew constantly tune front and rear wing angle to get the most cornering power.

ROTATING INERTIA 15

Most car enthusiasts know that reducing weight will increase acceleration with the same horsepower. What is less well-known is that if the weight, and its distribution, of the driveline components are reduced, the improvement in acceleration can be much greater than that which would be realized for just reducing the car weight. The weight and its distribution of a driveline component about its center of rotation is called its *rotating inertia*. This resistance to rotational acceleration is called rotating inertia, because its effect is seen when the parts are accelerated rotationally. The easiest way to picture this condition is to compare how hard it would be to rotate two objects that weigh the same, as shown in Figure 15-1.

Because of the distribution of weight around the rotational point, it would be much harder to quickly rotate the dumbbell in Figure 15-1 than it would be to quickly rotate the cannon ball. Both of these objects have the same weight, but the dumbbell has much greater rotating inertia. If your car had a flywheel that had most of its weight concentrated around the rim, it would have greater resistance to revving quickly than a flywheel that had its weight concentrated around its center.

Obviously, a lighter flywheel would be even better because it would not only weigh less, but it would also have less rotating inertia. Many enthusiasts are familiar with this concept, so it is interesting to look at an example to see just how much effect rotating inertia has on a car's ability to accelerate.

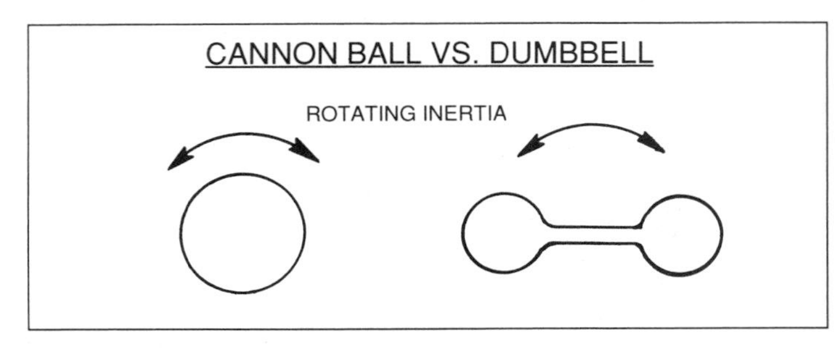

Figure 15-1. This comparison shows that it would be easier to rotate a cannon ball than it would to rotate a dumbbell. The reason for this difference is the dumbbell has a higher moment of rotating inertia.

As shown by the example worked out in the text, even a slight decrease in rotating weight can have a significant effect on acceleration.

VEHICLE ACCELERATION

Car enthusiasts know that their car will go faster if its weight is reduced. Less weight increases acceleration with the same available driving force. The formula for acceleration is:

$$\text{Acceleration} = \frac{\text{Force}}{\text{Weight}}$$

$$\text{Weight} = \text{Mass} \times \text{Gravity}$$

If we measure acceleration in terms of *g*-force, there is no need to convert weight into mass. For an example, let's find the acceleration of a 3000-lb. car that has 300 lbs-ft. of engine torque, a 5-to-1 gear ratio and rear tires with 13.5-in. static radius. We can find the force available to accelerate the car by:

$$\text{Force} = \frac{\text{Engine torque x axle ratio}}{\text{Rear tire static radius}}$$

$$\text{Force} = \frac{300 \text{ lbs-ft.} \times 5.0 \times 12 \text{ in.}}{13.5 \text{ ft.}}$$

$$\text{Force} = 1333 \text{ lbs.}$$

This assumes 100% driveline efficiency and no rotating inertia in the driveline. To find the acceleration of the car:

$$\text{Acceleration} = \frac{\text{Force}}{\text{Weight}}$$

$$\text{Acceleration} = \frac{1333}{3000}$$

$$\text{Acceleration} = .444\ g\text{'s}$$

This comparison shows that a V6 crankshaft will have less rotating inertia than a V8 crankshaft because it has less total mass and its distribution about the centerline of rotation is similar. For the same horsepower output, the V6 engine will be able to accelerate the car quicker.

VELOCITY

Velocity is equal to the acceleration multiplied by the time. If we assumed a constant acceleration for a period of five seconds, we could tell how fast the car was going at the end of the straightaway. For explanation, we have chosen a speed at the start of the straight of 64.00 mph @ 4000 rpm.

$$\text{Velocity} = \text{A} \times \text{T}$$

$$\text{Velocity} = \frac{.444\ g\text{'s} \times 32.2 \text{ ft.} \times 5 \text{ sec} \times 60 \text{ mph sec}}{g\text{'s} \times \text{sec}^2 \times 88 \text{ ft.}}$$

$$\text{Velocity Change} = 48.74 \text{ mph}$$

$$\text{Velocity at end of straight} = 112.74 \text{ mph}$$

If we reduce the weight of the car by 15 lbs (.5%), the same engine output would allow an acceleration of .446 *g*'s. This acceleration over the same five second period would give a speed at the end of the straightaway of 112.95 mph, an increase of .21 mph.

This photo of a Tilton clutch and flywheel shows the efforts made to reduce the rotating inertia on these parts that operate at engine speed. Space-age lightweight parts are used to reduce rotating weight.

CHASSIS ROTATING INERTIA

All the rotating parts in the chassis have inertia which resists angular acceleration. These parts include the tires, the wheels, the brake rotors, the hubs and the ring gear and the differential. To show the effect of the rotating inertia on these parts that turn at wheel speed, we used the same example car but with a 15-lb. reduction in these parts. Under the same test conditions, the 15-lb. reduction in rotating inertia would allow the car to accelerate to 113.34 mph for a .60 mph improvement. This shows that a 15-lb. reduction in rotating inertia on the chassis rotating parts will have three times the benefit of a 15-lb. weight reduction on the rest of the car. The 15-lb. reduction in chassis rotating weight was assumed to be a 12-inch diameter steel disk of constant section mounted on the rear axle drive flanges.

DRIVELINE ROTATING INERTIA

The rotating parts of the driveline include the crankshaft, the flywheel, the clutch, the transmission gears and the driveshaft. Since these parts operate at a much higher rpm, the effect their rotating inertia has on acceleration is much greater. If we reduced the rotating inertia of the flywheel on the

This Jerico transmission only has two forward gears. Removing gears that rotate at engine speed reduces the rotating inertia of the transmission, which permits the car to accelerate faster.

Even a driveshaft can be designed and built that has less rotating inertia. This type of refinement indicates that racers have found that reducing rotating inertia really improves acceleration.

example car by 15 lbs., it would allow the car to accelerate to 115.70 mph under the same test conditions for a 3.0 mph improvement. This shows that the effect of reducing rotational inertia on driveline parts has 15 times the benefit of just reducing the weight of the car. These examples show why there is so much emphasis by race car component manufacturers on reducing the rotational inertia of driveline parts. Actual racing experience has proven the benefits suggested by the above examples. The 15-lb. reduction in driveline rotating weight was assumed to be a 13-inch diameter steel disk of constant section mounted on the crankshaft flange.

Another way to put numbers on the effect of reducing rotating inertia is to compare the horsepower equivalents as shown in Chart 15-1. The important thing that this Chart shows is that it is 15 times as important to reduce the weight of components that rotate at engine speed than it is to reduce the weight of other components of the car. Components that rotate at axle speed are 3 times as effective as non-rotating components. ■

Chart 15-1

	Baseline Car Weight	Less 15 lbs. Total Weight	Less 15 lbs. Rotating Weight. @ Axle Speed	Less 15 lbs. Rotating Weight @ Engine Speed
Weight	3000 lbs.	2985	2985	2985
Acceleration Force	.444 g's	.446	.450	.474
Speed @ start of straightaway	64.0 mph	64.0	64.0	64.0
Speed @ end of straightaway	112.74 mph	112.95	113.34	115.70
Speed Change	—	.20	.60	3.00
Equivalent Horsepower	428.41	430.55	434.83	460.51
Horsepower Change	—	2.14	6.42	32.10
Improvement Factor	—	1	3	15

VEHICLE TESTING & TUNING 16

All of the design and build information we have available is based on many years of testing by thousands of racers, engineers and car enthusiasts. Testing shows us what works and what doesn't. As the testing of a particular aspect of chassis design indicates what is the best configuration, this information is then considered as the best approach. Over time, the best approach changes from one design to another. We call this progress. This changing base of knowledge tells us there is no right or wrong way to design a chassis. Each car, and its use, is a specific problem that needs a specific solution. Regardless of the design approach taken, the final judgment is based on how well the car performs.

The initial testing of any car often indicates areas that can be improved with modifications to the basic design. The development of these test-inspired ideas is called *chassis tuning.* Experience has shown that testing and tuning is more important to the overall performance of a car than the design direction that was taken. The following techniques are often used to test and tune a high-performance car.

WHEEL ALIGNMENT

It is impossible to do any meaningful testing or tuning unless you know that the front and rear wheels are aligned correctly. Here are some tips on how to align your car's wheels.

Setting Front-End Alignment

Many car enthusiasts regard alignment as a black art. The truth is, setting alignment is nothing more than pointing the wheels in the right direction. If you happen to have a mega-dollar rack at your disposal, you're home-free. But if you're like most enthusiasts, you have to improvise. Since you're doing the work yourself, on your own car, the results can be superior to the flat-rate workmanship sometimes encountered at uninspired alignment shops.

Figure 16-1. Now that the design and manufacture of your chassis is complete, you need to test it to make sure it works! It's an ongoing process that requires patience and careful analysis of the data.

Figure 16-2. You can use a protractor-level to get a fairly accurate measurement of your car's camber and caster. Measure the front wheels and the rear wheels to determine the camber angles of each wheel. Be sure the car is on level ground.

So how do you set your car to our specs when the alignment shop wants to use the factory setting? Simple. Do it yourself. Perfectly satisfactory results can be achieved with quite ordinary equipment.

Camber—To measure camber, a protractor level (available at any Sears store) works well. With the car parked on level ground and the front wheels straight ahead, place the level against the tire sidewall, avoiding raised lettering and other sidewall irregularities. Centering the level's air bubble will indicate the camber angle. For the negative camber, both front tires should tilt in toward the center of the car at the top (Figure 16-2).

Caster—Caster is the inclination of the front spindle. Positive caster—the rearward tilt of the top of the spindle—is essential for high-speed stability. A fairly accurate measurement of caster can be made by taking two camber readings. First, record the camber when the wheels are turning full right, then repeat the measurement with the wheel turning full left. The difference between the two readings closely approximates the caster angle. For the left-front wheel, positive caster is indicated when more negative camber is produced when the wheels are turned right. For the right-front wheel, positive caster is indicated when more negative camber is produced when the wheels are turned left. To increase caster, tilt the upper control arm rearward in the car.

Toe-In—Measuring toe-in is a snap. Two blocks of wood or oil cans, an 8-foot length of rectangular tubing, and a tape measure are all that's required. In a pinch, you can even substitute a straight 2 x 4 for the tubing. Set the tubing on top of the two oil cans as shown in Figure 16-3, with the tubing just touching the front tire sidewalls. Again, avoid raised letters which may disrupt the measurement. The oil cans are there to raise the straightedge above the tire bulge. Make sure the steering wheel is pointed straight ahead and be certain that the door is closed firmly. Sight or measure between the rocker panel and the straightedge. If the rocker panel and the tubing are parallel, the front wheels have zero toe-in. If the straightedge is closer at the front of the rocker panel than at the back, the front wheels are toed-in. Since the rocker is about three times as long as the tire diameter, the offset should be three times as great as the recommended toe-in specifications. Thus, if the front wheel has the desired 1/16-inch toe-in, the difference between your measurements from the straightedge to the front and rear of the rocker panel will be 3/16-inch. Toe-in is adjusted by turning the threaded sleeves which connect the inner and outer tie-rods ends.

Figure 16-3. A straight piece of tubing on a couple of oil cans can be used to get an accurate measurement of your car's front and rear toe-in. The cans are used to raise the straightedge above the tire bulge. Comparing the straightedge to the rocker panel eliminates the possibility of having the rear axle crooked in the car.

Setting Rear End Alignment

IRS—The techniques for measuring rear-wheel alignment are the same as those for measuring front-wheel alignment. Experience has shown that cars with independent rear suspension work best with between 1 and 2 degrees of negative rear-wheel camber. They also need between 1/16-in. and 1/8-in. of rear-wheel toe-in. The procedure for setting rear-wheel camber and toe-in can be obtained by consulting the service manual for your car. When setting toe-in, it is best to compare each wheel independently to the car rocker panel or door rather than to the other wheel. Using this procedure will prevent setting the car up with a "dog track" condition where the rear wheels won't follow in line with the front wheels.

Live Rear Axle—Rear-wheel camber and toe are as important on a car with a live-rear axle as they are on a car with an independent rear suspension. Since the rear axle is fairly rigid, it must be bent to change the rear-wheel camber and toe. One way to do this is to heat the axle tube red-hot so that it shrinks when it cools. This procedure, shown on page 81, can produce up to about 1 degree of negative camber and toe-in, which is what we recommend for most applications.

When setting the rear-wheel alignment on a car with a live rear axle, it is very important to be sure that the rear axle is in the car at 90 degrees to the car centerline. An easy way to determine this is to measure the toe-in of each rear wheel in reference to the car's door or rocker panels. If both of these measurements are the same, you know your rear axle is in the car squarely.

Tire temperatures can help determine the correct tire pressure, camber setting and give an indication on tire survival.

READING TIRE TEMPERATURES

Unless you understand how it's done, reading tire temperatures is a black art. The information that you can determine from reading tire temperatures includes:

Optimum tire pressure

Optimum camber setting

If the tire will survive

To a limited extent, the chassis set-up

The key to understanding tire temperatures is to recognize that if one section of the tire is working harder, its temperature will be higher. Tire temperatures are usually read on the inside, the middle and the outside of the tread. Readings should be made as soon as the car comes off the track, since the variations in temperature will normalize the longer the car sits still. Tire temperatures are measured with a pyrometer. When interpreting tire temperatures, you have to also take into consideration factors such as a banked track vs. a flat track; a long straight vs. a short straight; track temperature and driver skill. It is best to write down each temperature for each tire as shown in Figure 16-4.

All of the temperatures are recorded as if the viewer is looking at the car from above and from the driver's seat. The right front tire's temperature is logically at the right front and equally important, the outside tire temp is recorded at the outside edge of the tire. This procedure must be strictly followed, or you won't be able to make any sense of the data.

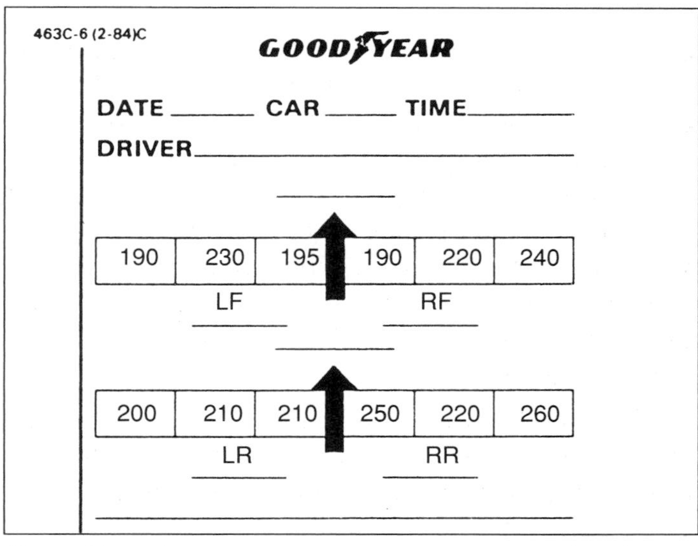

463C-6 (2-84)C

GOODYEAR

DATE ______ CAR ______ TIME ______

DRIVER ______

190	230	195	190	220	240
LF			RF		

200	210	210	250	220	260
LR			RR		

Figure 16-4. Tire temperatures and pressures must be taken as soon as the car comes off the track. Be careful to record each temperature in its correct location because a mix-up can lead to the wrong analysis. Goodyear Tire & Rubber will provide temp sheets like this, as long as you run their tires, of course. Courtesy Goodyear.

Here is a brief explanation of how you can use tire temperatures to optimize suspension settings.

Optimum Tire Pressure—If we accept that the section of the tire that works the hardest will run the hottest, too high an inflation pressure will be indicated by the center temperature being significantly hotter than the edge.

A significant temperature variation is usually above 10 F. On the example data sheet (Figure 16-4), the left front tire is obviously running at too high a pressure since the middle temperature is considerably hotter than either edge. Tire temperature distribution is not the only way to determine optimum tire pressure, but it is an important factor to consider. If we dropped the pressure in the left front of our example car, I would expect to see a more equal temperature distribution similar to 210-220-220 degrees. This type of analysis is the same for oval track cars as it is for road-racing cars.

Optimum Camber Setting—Tire temperatures can be used to optimize camber settings using the same logic as was used to optimize tire pressure. If the tire temperature for the right front tire is as shown on the sample data sheet, the car does not have enough negative camber. This can be seen by the fact that the outside edge is running much hotter than the inside or the middle. If the outside is running hotter, it's working harder, so there is not enough negative camber. If more negative camber was adjusted into the right front, all of the tire would share the load equally, and the tire temperature distribution would look more like 210-230-230 degrees. This change in camber setting would also probably result in faster lap times, and less push, since the right front tire would be working at a higher efficiency. A similar analysis is used to set up a road-racing car or an oval track car. In either case, the temperatures of the tires on the outside of the turn are most important.

Tire Survival—Most racing tires work best in the 200—250 F. range. If a tire sees temperatures above this level in practice, they are sure to fail under the continued loads that are seen during the race. If we observed tire temperature, as shown in Figure 16-4, at the right rear of the example car, something would have to be changed to prevent burning up the tire. Chances are the tire is already gone if it has seen this level of temperature, even for a few laps.

Chassis Set-Up—Since the right rear on the car is running hot, I would ask if the driver has noticed if it was spinning coming off the corners. This can be a problem on short tracks with very lightweight and powerful cars. If there is no other remedy, such as a larger tire or a harder compound, the driver might have to modulate the throttle a little to keep from smoking the right rear tire. An overly hot right rear tire can also be caused by an oversteering chassis set-up. If too much of the car's cornering force is being generated by the right rear tire, it will run hot. The driver should obviously notice this condition to confirm the analysis of the chassis set-up. If you set up the car for more understeer, the rear tire temperature will drop and lap times should improve.

The left rear tire on the sample tire data sheet shows the type of temperature distribution that is normal and desired. The inside edge is slightly cooler because the car never leans on it. The outside is at the same temperature as the middle so the tire is working near its maximum efficiency.

Portable scales can help you determine individual wheel weights to determine static loads, which will help you balance the tires' vertical loads during cornering.

BALANCING UNDERSTEER & OVERSTEER

We know from our understanding of a tire's cornering efficiency (Chapter 1) that its cornering power can be changed by adjusting how much vertical load it sees during cornering. Making a car corner at the highest possible speed requires that all the tires work at their highest efficiency. This means then that you need to balance the tire's vertical loads during cornering. From prior chapters, we know that the following chassis elements can be adjusted to accomplish this.

Achieving an acceptable balance between oversteer and understeer can only come from hours of testing and tuning. Experimenting with different front and rear stabilizer bars, and different spring rates, is the first step. Photo by Michael Lutfy.

Stabilizer Bars

In Chapter 3, we showed how using a front and/or rear stabilizer bar can affect the front-to-rear distribution of the lateral weight transfer during cornering. When a front stabilizer bar was added to the example car, its resistance to body roll caused more of the lateral weight transfer from cornering to be absorbed by the outside front tires. The amount of lateral weight transfer absorbed by the front tire was then more than it would see from just the weight on the front of the car. Changing the roll stiffness of the front and/or rear of the car can be used to alter

If the driver comes in and says the car is understeering going into a corner, you can try to add a more effective rear stabilizer bar and/or lower the front spring rate. Photo by Michael Lutfy.

how much of the total lateral weight transfer can be distributed between the front and the rear outside tires.

Because front and/or rear stabilizer bars can be used to change the roll stiffness of the front and/or rear of a car, they are quite effective in changing the understeer/oversteer characteristics. A more effective front stabilizer bar will cause more understeer, because it causes the outside front tire to absorb more of the total lateral weight transfer. If the outside front tire carries more vertical load, its cornering efficiency will be reduced and the front-end will lose cornering power.

Spring Rates

In addition to front and/or rear stabilizer bars, changing spring rates can also be used to alter the understeer/oversteer characteristics. Higher spring rates at the front will increase the front roll stiffness vs. the rear roll stiffness. As was the case with the front stabilizer bar, more front roll stiffness causes more of the lateral weight transfer to be absorbed by the outside front tire, which results in more understeer. Stiffer rear springs have the opposite effect, because they make the rear outside tire absorb more of the total lateral weight transfer. When the rear tires are asked to carry more weight, their cornering efficiency goes down and the car will have more oversteer. Since the tires' cornering efficiency is very sensitive to vertical loading, even small changes in front or rear spring rates can have a significant effect on whether a car understeers or oversteers.

Front & Rear Roll Center Height

If the roll center of the front suspension is raised, it has the same effect as increasing the front roll stiffness. This means that more of the lateral weight transfer will be absorbed by the outside front tire and the car will understeer more. Raising the front roll center height also reduces the amount of camber gain, which causes even more understeer. The combined effects of more roll stiffness, and less camber gain, can cause a considerable loss in front-end cornering power.

Frame Stiffness

If you want to alter the roll force distribution of your car to fine-tune its understeer/oversteer characteristics, you can do it most easily by changing

If the car oversteers coming out of a turn, the driver won't be able to get on the throttle as quickly, which is especially important when coming onto a straightaway. To reduce corner exit oversteer, start by using a less effective rear stabilizer bar, and/or use higher front or lower rear spring rates. More anti-squat is a good solution to this problem. Photo by Michael Lutfy.

the diameter and/or swing-arm lengths of the front and/or rear stabilizer bars. In order for these changes to have a significant effect on your car's handling, the frame stiffness must be good enough to transmit the load from the front of the car to the rear of the car. This requirement is called torsional stiffness, which is discussed in Chapter 12. If you find little change in handling when you change stabilizer bars, the likely problem is inadequate torsional frame stiffness.

APPLICATIONS

All of the above information can be applied to any given car to determine how best to improve its handling and cornering speed. Because the application of all of these elements can become complex, here are some examples of applying the information to a real car with real handling problems.

Corner Entry Understeer

A common complaint from race car drivers is, "The car understeers at the entrance of a turn. What can I do to improve this condition?"

Before getting into the specific things that can be done to improve corner entry, it is best to try and analyze why a car wants to understeer on corner entry. When a car enters a corner, there is some weight transfer from the rear of the car to the front of the car. Even if the brakes aren't used, just backing off the throttle causes deceleration, which transfers weight forward. The transfer of weight forward and the transferring of weight from the inside tires to the outside tires (due to the cornering loads) causes an increase in the outside front tire loading. As shown by the tire curve in Chapter 1, increased tire loading causes a loss in tire cornering efficiency, so the front-end of the car will have less cornering power than the rear-end of the car. Less cornering power at the front causes understeer. Also, if the driver uses the brakes on corner entry, some of the tire's total traction will be used to decelerate the car. Since this leaves less traction for cornering, there will be even less cornering power available and the car will want to understeer even more.

From the above analysis, you can see that a car's turn-in performance will improve if the weight on the outside front tire is reduced during cornering. Here are some ways to reduce the load on the outside front tire during cornering.

More Effective Rear Stabilizer Bar—Since we want to transfer some of the cornering load from the front outside tire to the rear outside tire, a more effective rear stabilizer bar will help. A less effective front stabilizer bar will also help, but this will cause an increase in roll angle, which leads to more understeer. This is obviously not the preferred option.

Lower Rate Front Springs—Since the front springs contribute to the roll stiffness, and therefore, the amount of roll force the front tires absorb, a lower rate front spring will result in less understeer. Many people have trouble with this recommendation, because they think a 400 lbs-in. spring will handle more load than a 300 lbs-in. spring. With our understanding of tire characteristics and vertical loading, we know this is a false concept. Another option is to increase the rear spring rate, which will also decrease understeer.

Corner Exit Oversteer

Another common complaint from race car drivers is, "The car oversteers at the exit of a turn. What can I do to improve this condition?"

Experienced racers know that how well a race car hooks up on turn exit is the key to better lap times. The sooner the driver can get the throttle wide open, the faster the car will get around the track. Because of lateral weight transfer during cornering, the outside rear tire is the primary element in getting off a corner quickly.

An analysis of why race cars want to oversteer on corner exit might be a good way to find ways to solve this problem. When a race car exits a corner there is weight shifted from the front of the car to the rear of the car because of acceleration. Since there is also weight transfer from the inside tires to the outside tires, the outside rear tire will see more weight than it would under static conditions.

From the tire performance curve, you know that an increase in loading will cause a reduction in cornering efficiency. And from the Circle of Traction concept, we know that if some of the rear tire's total traction capability is used to accelerate the car, there will be less available for cornering loads. Both of these effects result in less cornering power available from the right rear tire, so the car will oversteer. If you reduce the dynamic weight on the outside rear tire, it will improve its cornering efficiency, and there will be more rear cornering power, so the car will not oversteer as much during corner exit.

If you have corner entry understeer and corner exit oversteer, the best solution is to set the spring rates and stabilizer bars for neutral steer, which is perfect cornering balance in the middle of the turn. Dialing in more anti-squat and more rear aerodynamic downforce will help reduce corner exit oversteer. Photo by Michael Lutfy.

Here are some ways to reduce the load on the outside rear tire during cornering.

Less Effective Rear Stabilizer Bar—Since we want to reduce the load on the rear outside tire during cornering, a less effective rear bar is needed. A more effective front bar, of course, will also help this condition.

Higher Front or Lower Rear Spring Rates—Both of these changes will have the same effect because they will transmit more of the roll force to the front outside tire away from the rear outside tire.

Balance

The above indicates that what helps corner entry problems hurts corner exit problems and vice-versa. This is the balance problem most race car drivers experience. A typical complaint is "The car understeers on turn entry and oversteers on turn exit. What can I do to improve both conditions?"

This combination of problems results from the tire loading changes that occur when a car goes from turn entry to turn exit. We have found that the best solution is to set the spring and stabilizer bars for perfect cornering balance in the middle of the corner. If the car has no understeer and no oversteer in the steady-state cornering condition, this will provide the highest cornering speed. This is called *neutral steer*.

If a car is set-up to have neutral steer in the middle of a corner, it will also be reasonably good for turn-in at the corner entrance. If we can improve the corner exit performance of a neutral steer car, the result will be excellent handling. The following are some suggestions to achieve this.

Anti-Squat—From our knowledge of the anti-squat feature of a rear suspension, we know that it will increase the loading between the rear tires and the pavement during acceleration. Since we want to accelerate at the corner exit, this temporary increase in vertical load will provide the outside rear tire with extra cornering power to give the car the forward thrust without losing any of the cornering power it needs to keep the car from oversteering. The application of anti-squat designs and adjustments are the primary means of getting a neutral steer car to hook up and accelerate out of a corner.

Aerodynamic Downforce—Since aerodynamic downforce can be used to increase the tire vertical loads with the resulting increase in cornering loads, balancing the front-to-rear aerodynamic downforce can also be used to tune a car's understeer/oversteer characteristics. More aerodynamic downforce at the rear of the car will increase the rear wheel loading and the cornering power of the rear tires. More cornering power at the rear reduces oversteer, so this technique can also be used to help a car accelerate out of a corner. Obviously, the effects of aerodynamic downforce will not help much at low speeds.

MEASURING G-FORCE

The forces acting on a car are measured in terms of *g*-force. If a car is cornering at 1.0 *g*'s, the tires are providing a cornering force that is equal to the weight of the car. If a car is accelerating forward at .50 *g*'s, the engine is providing enough force at the rear tires to push on the car with a force equal to one-half of the weight of the car.

Since *g*-forces give us a good way to measure race car performance (around corners as well as for acceleration and braking), it would be helpful if we could measure them while the car was actually running. The aircraft industry has had devices to measure *g*-forces for many years. But, since their equipment is very expensive, not too many enthusiasts have been able to use it. A less expensive means of measuring and recording g-forces is provided by a device called the *g-Analyst*.

The *g*-Analyst provides a high degree of accuracy and versatility in a small, rugged package. Automotive enthusiasts, and especially racers, can now measure and record the *g*-forces their car experiences on the track (Figure 16-5).

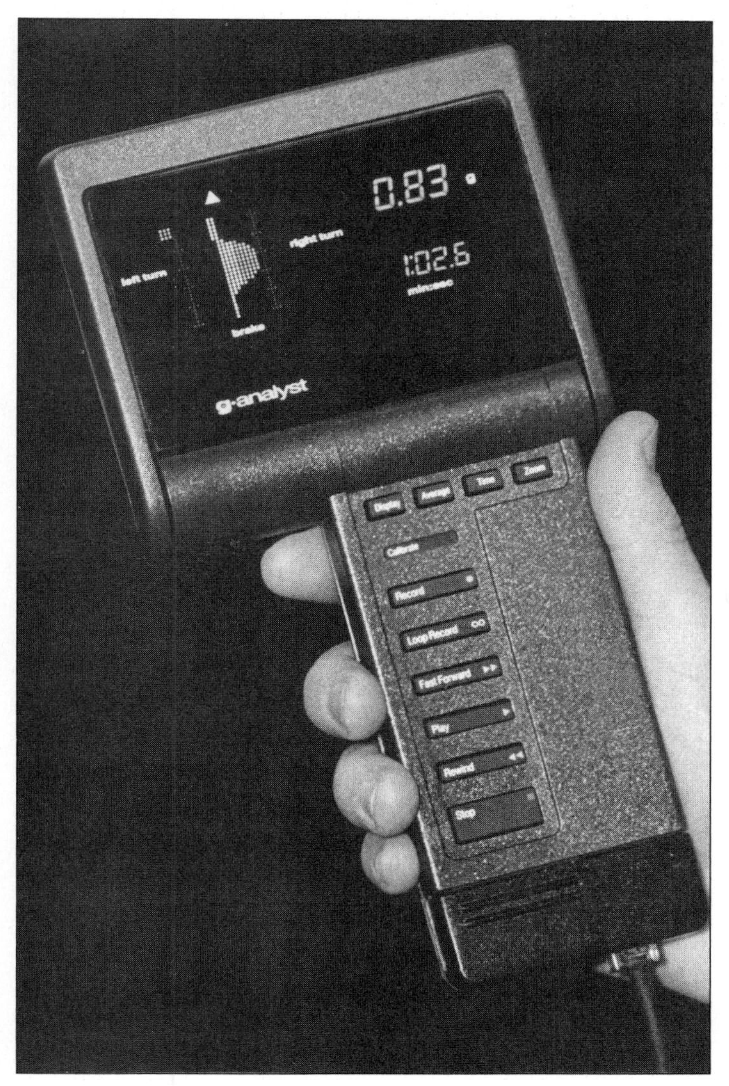

Figure 16-5. The g-Analyst consists of a display head, the transducer, and the connecting wires. A 12-volt power source, like a cigarette lighter, is required.

The *g*-Analyst

The *g*-Analyst is basically a *g*-meter that is married to a tape recorder. The g-meter measures the *g*-forces and the recorder provides an eight-minute tape of these measurements. After recording the *g*-forces that your car experiences on the road or track, you can playback the data when the car is stopped. The *g*-Analyst includes a small screen which can be used to display the test results. These results can be analyzed individually (lateral *g*'s are cornering power and fore-and-aft *g*'s are acceleration and braking) or together to show the Circle of Traction.

For a more detailed analysis, the *g*-Analyst can be combined with a personal computer and a printer to provide a data presentation as shown in Figure 16-14.

The *g*-Analyst is easy to install in any car that has a 12-volt power supply (Figure 16-6). It is also easy to calibrate and to operate, so most enthusiasts should be comfortable using it.

Using the Data

With a *g*-Analyst, an enthusiast can collect the minimum of useful data and then analyze it quickly. If it is used in a test session, more specific information can be recorded that can be analyzed in depth. After a little experience, it is possible to quickly review the *g*-Analyst screen to get an overview of how a car is performing. Whenever the driver turns the steering wheel, applies the throttle or the brakes, he causes *g*-forces to be applied to the car. Since the *g*-Analyst measures and records these forces, it can be used to optimize a car's performance.

Measuring Acceleration Forces—The *g*-Analyst measures how well a car accelerates and records that data in the form of *g*-forces. If a car has a .50 *g*-force applied at the rear wheels, it will accelerate at a rate equal to one-half of the pull of gravity. In engineering terms this is described as:

$$g\text{-Force} = \frac{\text{Driving Force}}{\text{Weight}}$$

The *g* of acceleration increases as the driving force goes up and/or the weight goes down.

The force that causes the acceleration comes from the engine and is applied to the car at the driving wheels. Therefore, you need to know your car's weight, peak torque, transmission and rear axle ratios, and the rear tire radius. The following is an example of how this power is transferred into an accelerating force. Let's assume our car has the following specifications:

Car Weight:	**3000 lbs.**
Peak Engine Torque:	**400 lbs-ft.**
Transmission Ratio:	**1.31**
Rear Axle Ratio:	**4.11**
Rear Tire Radius:	**13.0 inches**

If the peak engine torque is 400 lbs-ft. going into the transmission, we will have 1.31 times this amount coming out of the transmission (400 x 1.31) or 524 lbs-ft. With a 4.11 axle ratio, the torque at the rear wheels would be 2154 lbs-ft. (524 x 4.11) or 25848 in-lbs. At a static tire radius of 13.0 inches, the force applied by the tire to the pavement would be 1988 lbs. This is the force that causes the car to accelerate.

$$g\text{'s} = \frac{1988}{3000}$$

$$g\text{'s} = .66$$

The above example assumes 100% efficiency for the transmission, clutch, driveshaft and rear axle. In the real world this is never the case. A transmission in high gear might be 99% efficient. In the intermediate gears it might be only 97% efficient. Rear axle efficiency is dependent on the set-up and the type of axle. The ring-and-pinion in a Quick Change rear

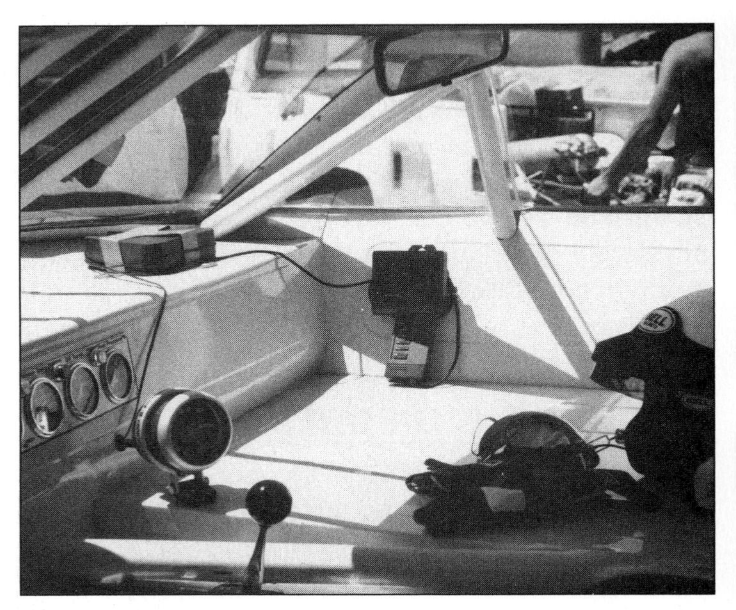

Figure 16-6. Mounting a g-Analyst in a car is fairly simple. The transducer must be mounted flat and in line with the car's centerline as described in the operation manual.

axle is more efficient than in a Ford 9-inch axle, but there is the added loss of the transfer gears. Rear axle overall efficiencies usually range from 90% to 95%. In addition to the losses experienced through the drivetrain, the actual car acceleration rates will be less, because all the rotating components have rotational inertia, which absorbs some of the power before it gets to the rear wheels.

With a *g*-Analyst, you can compare the maximum *g*-forces during acceleration to determine if your engine is running up to par. When comparing data from one event to the other, you will need to correct for transmission ratios, axle ratios and tire radii.

Because the *g*-Analyst records data every 1/10 of a second, it can also be used to measure your engine torque curve. Peak torque only occurs at one rpm level, but because the *g*-Analyst measures continuously, the *g*-forces it shows during acceleration are directly related to the torque output from your engine. It is possible for the driver to trigger the *g*-Analyst at any point during the acceleration to mark the screen so you can read the *g*-forces at any specific engine rpm (Figure 16-7).

Measuring Braking Forces—Since the *g*-Analyst measures forces in all directions all the time, you can read the amount of braking forces your car experiences with the same data record you used to read the acceleration forces. The data from a *g*-Analyst can also be helpful in training a new driver because

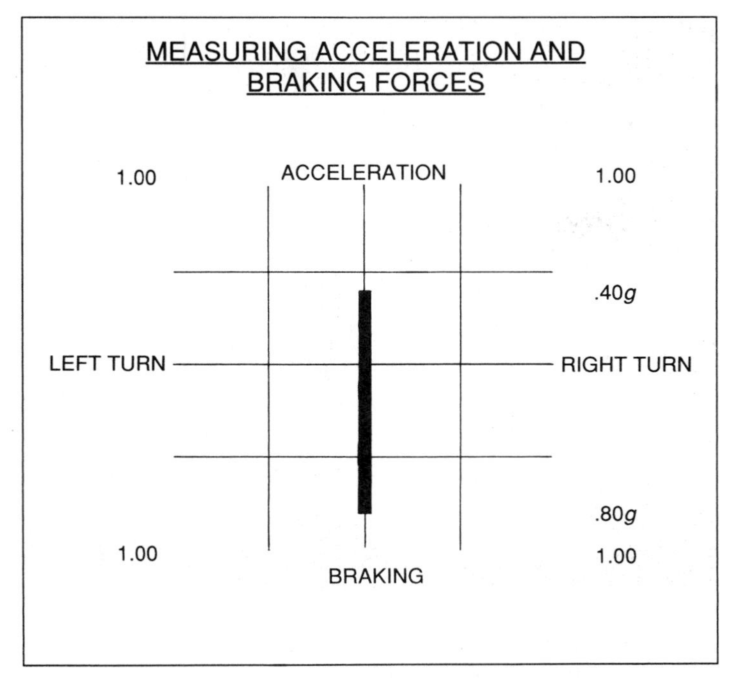

Figure 16-7. This print-out of a g-Analyst display shows a car accelerating at .40 g's and braking at .80 g's. The print-out can be extended over a period of time to show how the acceleration and braking forces change as the car is driven around the track.

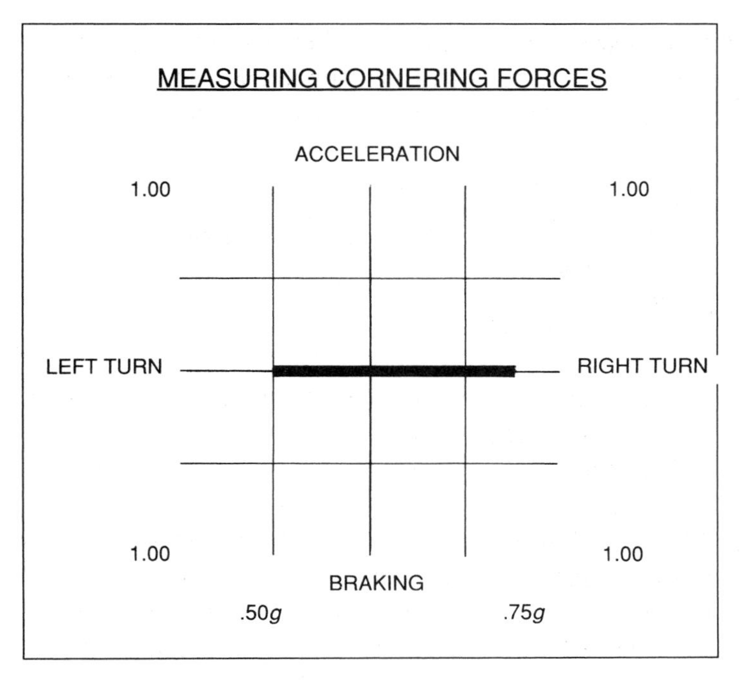

Figure 16-8. This print-out shows a car coming to the left at .50 g's and to the right at .75 g's. Again, this type of display can be seen over a period of time to show how a car generates cornering force throughout a corner.

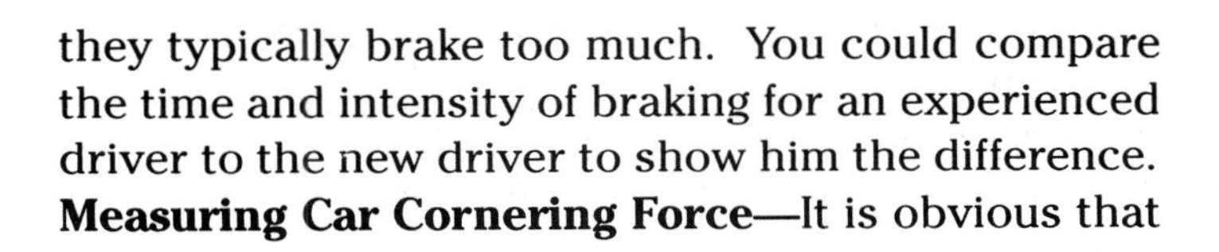
they typically brake too much. You could compare the time and intensity of braking for an experienced driver to the new driver to show him the difference.
Measuring Car Cornering Force—It is obvious that we will see higher *g*-forces during cornering if the car is set-up correctly and the tires are working to their maximum. This is the most basic data that a *g*-analyst can provide about cornering forces. The

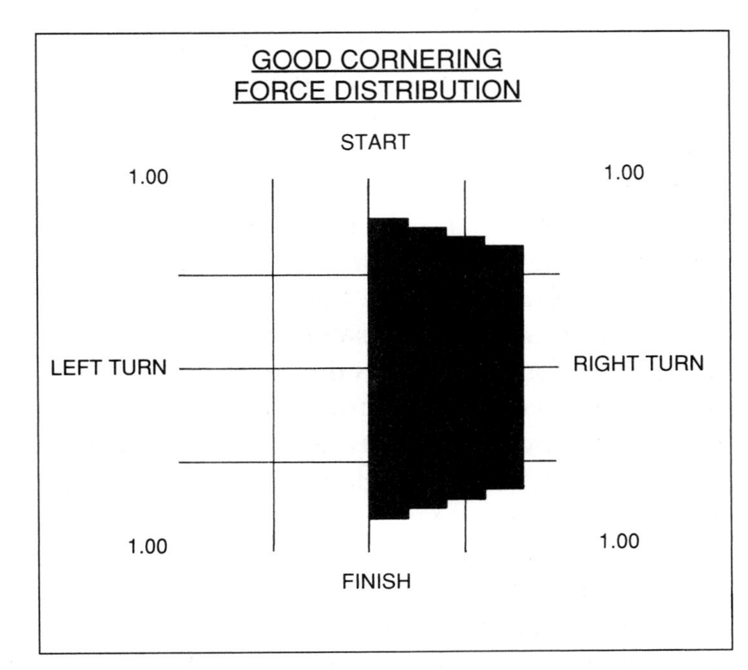

Figure 16-9. Here is a print-out of a car coming to the right throughout a typical corner. The maximum g-forces are .80, but the car reaches this level quickly and holds it for most of the corner. This force is considered good, because it shows the car and the driver are able to reach and hold the maximum cornering force.

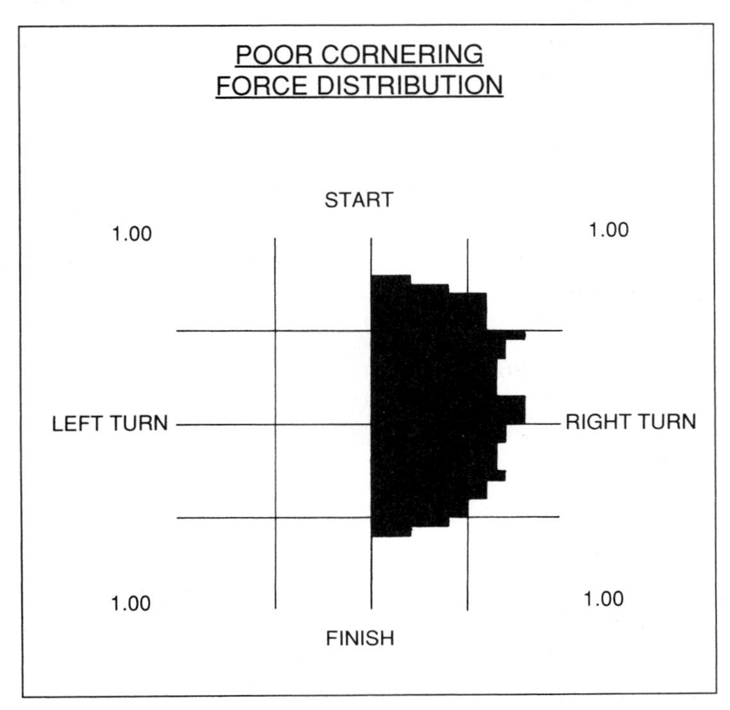

Figure 16-10. This print-out shows a car and driver that have trouble maintaining full cornering power throughout the corner. Either the car or the driver is having a problem.

readings will be dependent on the car set-up and the tires as well as the bank angle of the track and the condition of the track. You can switch from one type, or compound, of tire to another and your *g*-Analyst will tell you which provided the most cornering force. More importantly it can also tell you how much better one is from the other. Chassis set-up changes are often so small that they won't show up as big changes in maximum *g*-forces. With experience in using a *g*-analyst, it is possible to see changes in car set-up, but it is rarely a difference in maximum cornering force that shows up. You can use a *g*-Analyst to compare cornering forces from one track to the other, but you will need to correct for the bank angle to get meaningful results. Track conditions such as temperature, dust, moisture and the amount of rubber on the track can also have a big effect on your cornering force readings, so you will have to use some judgment in making the evaluations (Figure 16-8).

Cornering Force Distribution During A Corner— Since the g-Analyst allows us to collect and record the *g*-forces throughout a corner, it provides us with enough data to see how the g-forces vary as the car goes around the corner. A typical corner display, like shown in Figure 16-9, would be displayed on the *g*-Analyst screen.

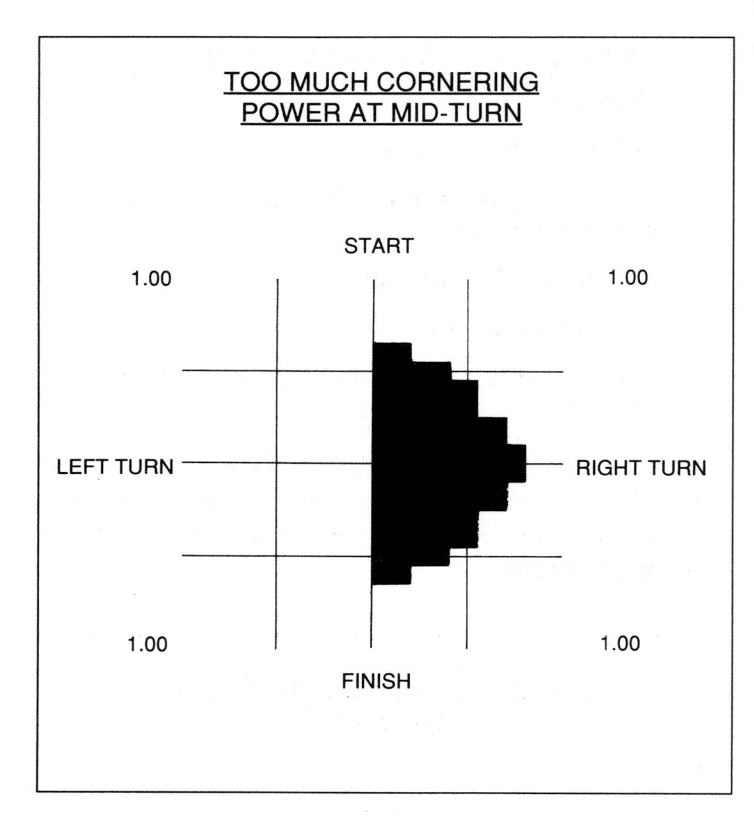

Figure 16-12. A print-out like this one could indicate that the driver is not using all of the car's cornering power all of the time. But, it could also indicate a steeper bank in the track at mid-point, so you have to use good judgment in interpreting the data. Drivers don't like to be criticized.

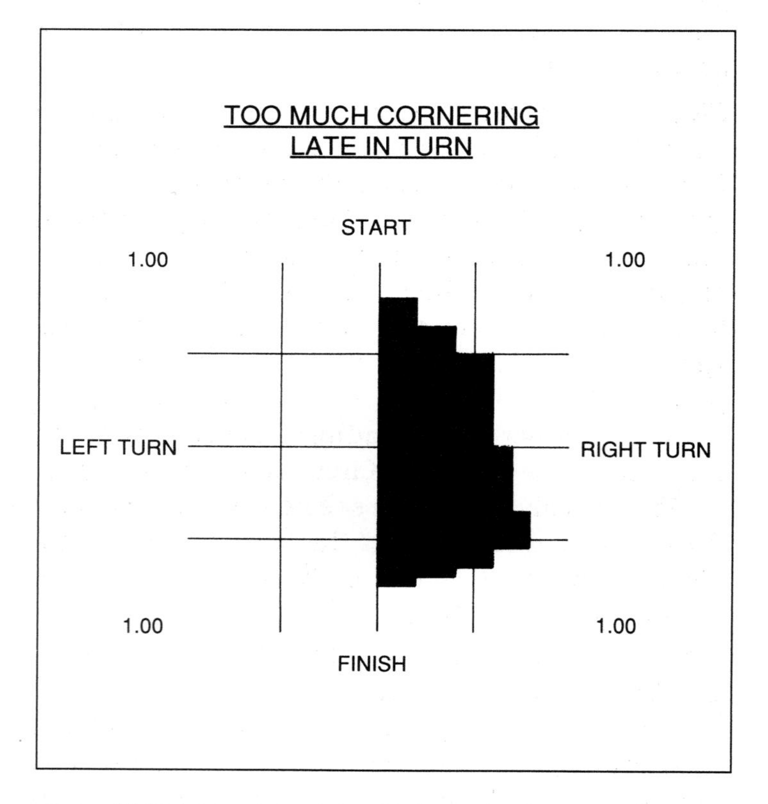

Figure 16-11. This print-out clearly shows the driver is using more cornering power at the end of the turn. This would indicate that he is taking too early of an apex and running out of road on the exit. It could indicate that the driver is timid about pushing the car to its limits as soon as the turn is entered.

This would be a pretty good corner, because the data shows that the car and driver were able to quickly achieve maximum cornering power and hold it through the corner. The data not only shows good cornering force, it also shows that the driver can control the car at maximum cornering force for the entire corner.

If the *g*-Analyst screen display for a corner looks like Figure 16-10, we would see some indication of problems. In this display it is obvious that the driver cannot control the car at maximum cornering force throughout the corner. It could be a driver problem, but if the driver is experienced, it is probably a car set-up problem. If the screen display looks this irregular, the driver has probably already complained. If he hasn't, show him the display and see if you can determine what might need to be changed.

Some of the likely problem areas would be:

Oversteer

Not Enough Anti-Squat

Loose Suspension or Steering Bushings or Connections

Too High a Spring Rate

Too Firm Shocks

Poor Suspension Geometry or Steering Geometry

If there is a severe bump in a given corner it will cause the cornering forces to vary somewhat. Here again, you will need to use some judgment in making your evaluations.

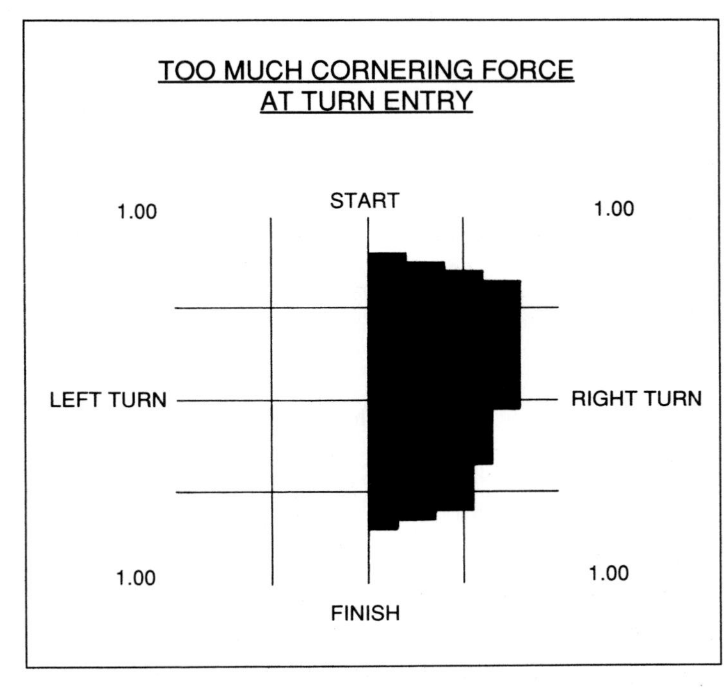

Figure 16-13. A print-out like this could indicate that the driver is throwing the car into the entrance of the turn and scrubbing off too much speed. The lack of speed throughout the remainder of the turn would result in the lower g-forces at the exit.

Evaluating Drivers—The *g*-Analyst can also be useful in measuring different driver techniques. Even experienced drivers can learn from other experienced drivers and comparing their display images offers a way to do this. Two different drivers could have the same total time through a corner, yet one of them might be faster on the entrance while the other could be faster on the exit. You have to be careful not to hurt any egos, but think how fast your car would be if you could have both the faster entry speed of the one driver and the faster exit speed of the other driver.

The following are some examples of different driving techniques as they would show up on the *g*-Analyst display screen.

Figure 16-11 shows a driver who is using more cornering power late in the corner. This indicates that he is starting to turn-in too soon and leaving most of the turn until last. Some tracks or traffic conditions might require this technique, but this is not usually the fastest way around a corner since the power cannot be applied until it's too late.

Figure 16-12 shows the driver using most of the cornering power in the middle of the corner. Again, track or traffic conditions might require this technique, but it is probably slower than if the maximum cornering power was used for the entire length of the corner.

Figure 16-13 shows the *g*-forces that a car would see if the driver came in hard and used maximum cornering power at the entrance and through the mid-portion of the corner. After the mid-portion, the *g*-forces would taper off, because the driver would be using more of the tires' total traction to start his acceleration early.

Circle of Traction—In Chapter 1, we discussed the force distribution characteristics of a tire called the Circle of Traction. Because the *g*-Analyst can measure and record g-forces during cornering and acceleration and braking all at the same time, it can also display the data at the same time. This is a useful tool to have, because only a small portion of a lap is spent in cornering, or in pure acceleration, or in pure braking. Usually a race car is both braking and cornering, or both accelerating and cornering at the same time. By using the Circle of Traction display on your *g*-Analyst, you can see how well the car and driver are able to use all of the tires' traction capability. An example of an effective use of a tire's full capability is shown in Figure 16-15.

As can be seen, the combination of the cornering and braking forces makes an almost perfect quadrant of a circle. The combination of cornering and acceleration forces are cut off at the top, because the engine doesn't have enough power to accelerate the car up to the point of wheel-spin. (Note: example shows cars only turning left on a circle track.)

Variations from this ideal Circle of Traction dis-

COMPUTER PRINTOUT

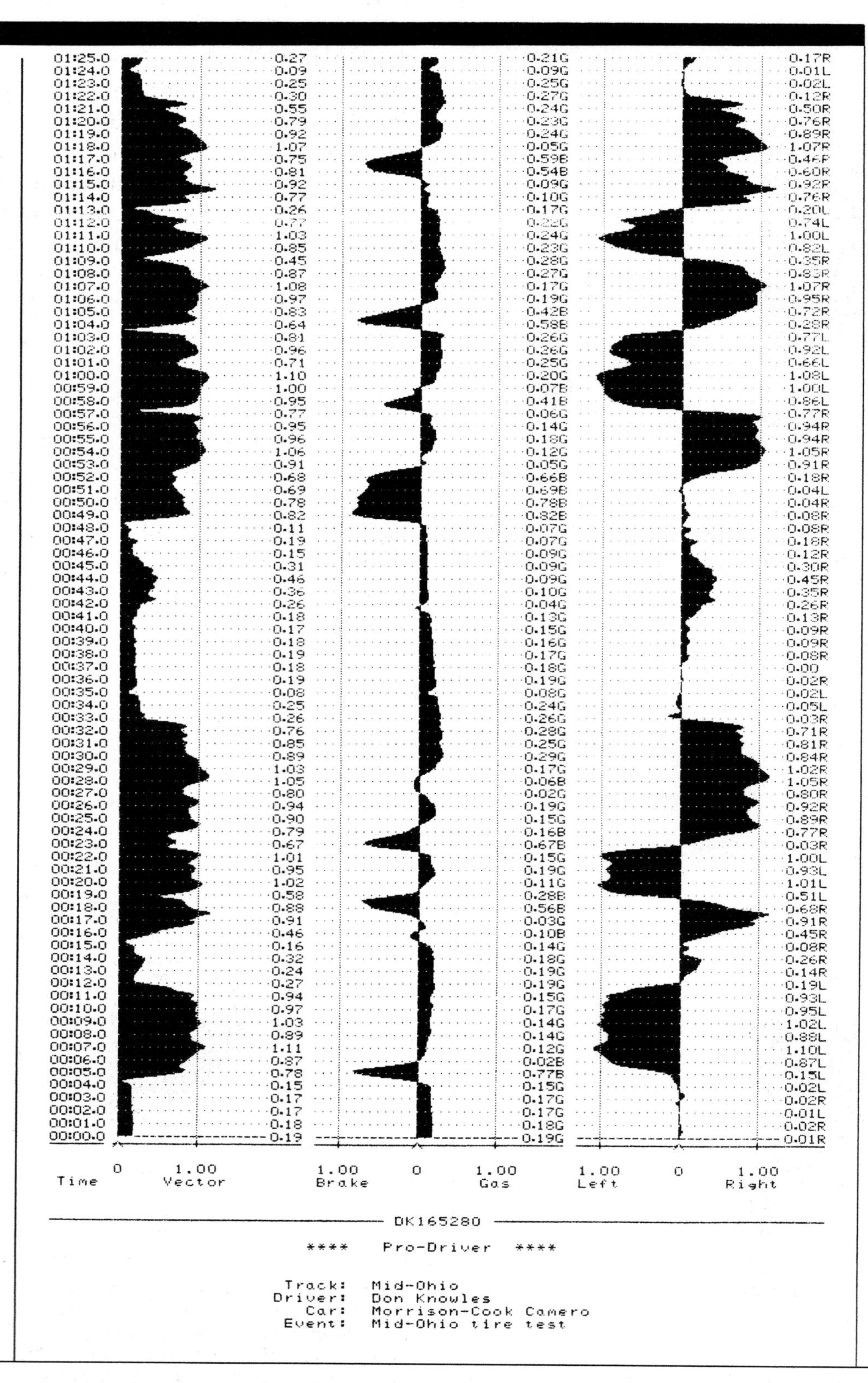

Figure 16-14. Instead of recording the g-force readings on the display screen, they can be processed on a personal computer to provide a diagram like this. This is one lap of cornering force readings at the Mid-Ohio road-racing track in Lexington, Ohio. Courtesy of Valentine Research, 10280 Aliance Road, Cincinnati, Ohio, 45242. Makers of the g-Analyst.

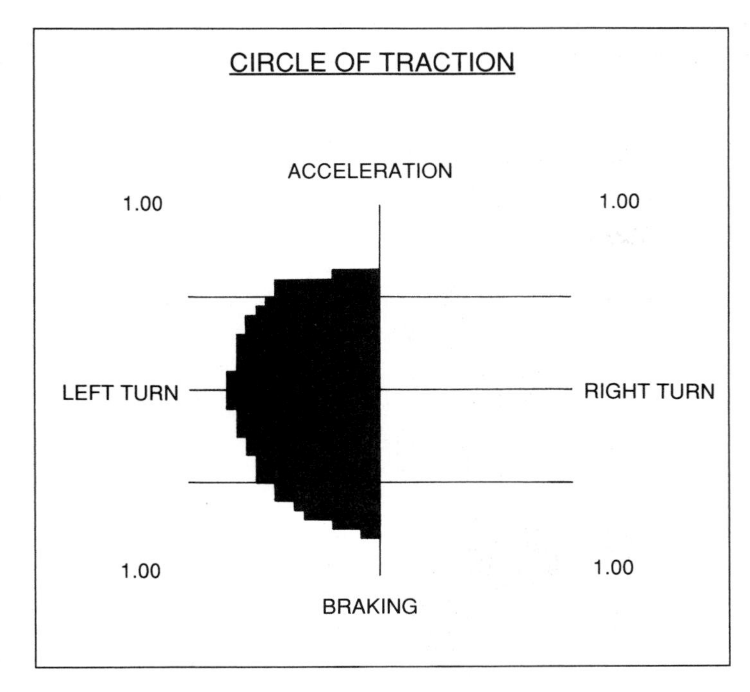

Figure 16-15. This print-out shows the cornering force and the acceleration-braking forces overlaid so you can see how well the driver and car utilize the full traction capability of the car. Only the left turn is shown because this is a circle track car.

play can indicate car and/or driver problems. If the Circle of Traction looks like Figure 16-16, there is a reason why the driver does not use the full traction capability of the tires.

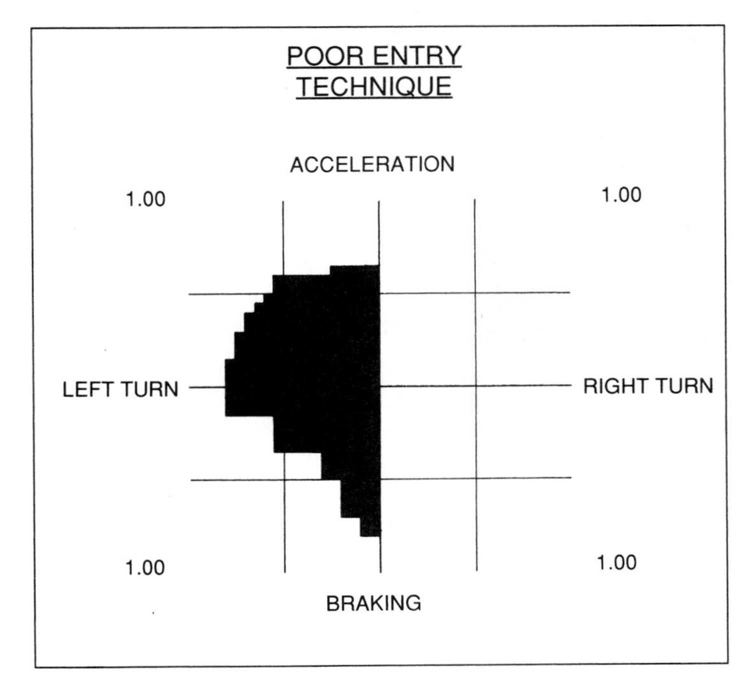

Figure 16-16. When a print-out shows heavy braking but not cornering force, it usually indicates a driver who is too careful. It could also indicate a car that has oversteer under braking, which could be corrected by limiting some of the rear brake bias.

When this happens it is usually because of an inexperienced driver or a car set-up that doesn't let the driver brake and turn-in at the same time.

In Figure 16-17, we see a similar problem on corner exit where the driver should be able to accelerate hard while still cornering. This type of Circle of Traction display would indicate that the driver was not able to get full throttle until after the car was already going down the straightaway. Again, the cause of this problem could be either an inexperienced driver or a car set-up that did not permit the driver to apply power early. One of the usual fixes for this problem in chassis set-up is to use more anti-squat.

Summary

The use of a *g*-analyst is a good way to get a measurable and readable record of the *g*-forces acting on your car. This information is helpful in understanding how these forces affect your car's handling characteristics. It also allows you to determine the level of cornering power and acceleration your car can achieve. ■

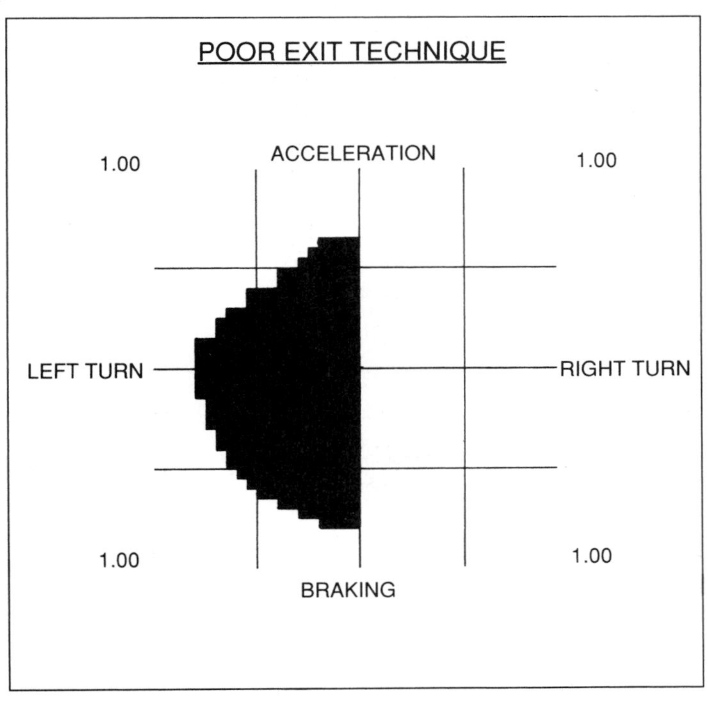

Figure 16-17. This print-out shows that the car and driver are not able to get full throttle while cornering hard. This will be a slow race car unless the condition is improved. If the driver is giving it his full ability, there must be something wrong with the car. A very likely area to look for an improvement would be to adjust the rear suspension for more anti-squat. More anti-squat will allow the driver to use full throttle sooner and this will show up as a fuller Circle of Traction print-out on a g-Analyst.

INDEX

D

E

F

G

S

T

U

V

W

X

Z

NOTES

NOTES

OTHER BOOKS FROM HPBOOKS AUTOMOTIVE

HANDBOOKS

Auto Electrical Handbook: 0-89586-238-7
Auto Upholstery & Interiors: 1-55788-265-7
Brake Handbook: 0-89586-232-8
Car Builder's Handbook: 1-55788-278-9
Street Rodder's Handbook: 0-89586-369-3
Turbo Hydra-matic 350 Handbook: 0-89586-051-1
Welder's Handbook: 1-55788-264-9

BODYWORK & PAINTING

Automotive Detailing: 1-55788-288-6
Automotive Paint Handbook: 1-55788-291-6
Fiberglass & Composite Materials: 1-55788-239-8
Metal Fabricator's Handbook: 0-89586-870-9
Paint & Body Handbook: 1-55788-082-4
Sheet Metal Handbook: 0-89586-757-5

INDUCTION

Holley 4150: 0-89586-047-3
Holley Carburetors, Manifolds & Fuel Injection: 1-55788-052-2
Rochester Carburetors: 0-89586-301-4
Turbochargers: 0-89586-135-6
Weber Carburetors: 0-89586-377-4

PERFORMANCE

Aerodynamics For Racing & Performance Cars: 1-55788-267-3
Baja Bugs & Buggies: 0-89586-186-0
Big-Block Chevy Performance: 1-55788-216-9
Big Block Mopar Performance: 1-55788-302-5
Bracket Racing: 1-55788-266-5
Brake Systems: 1-55788-281-9
Camaro Performance: 1-55788-057-3
Chassis Engineering: 1-55788-055-7
Chevrolet Power: 1-55788-087-5
Ford Windsor Small-Block Performance: 1-55788-323-8
Honda/Acura Performance: 1-55788-324-6
High Performance Hardware: 1-55788-304-1
How to Build Tri-Five Chevy Trucks ('55-'57): 1-55788-285-1
How to Hot Rod Big-Block Chevys:0-912656-04-2
How to Hot Rod Small-Block Chevys:0-912656-06-9
How to Hot Rod Small-Block Mopar Engines: 0-89586-479-7
How to Hot Rod VW Engines:0-912656-03-4
How to Make Your Car Handle:0-912656-46-8
John Lingenfelter: Modifying Small-Block Chevy: 1-55788-238-X
Mustang 5.0 Projects: 1-55788-275-4
Mustang Performance ('79–'93): 1-55788-193-6
Mustang Performance 2 ('79–'93): 1-55788-202-9
1001 High Performance Tech Tips: 1-55788-199-5
Performance Ignition Systems: 1-55788-306-8
Performance Wheels & Tires: 1-55788-286-X
Race Car Engineering & Mechanics: 1-55788-064-6
Small-Block Chevy Performance: 1-55788-253-3

ENGINE REBUILDING

Engine Builder's Handbook: 1-55788-245-2
Rebuild Air-Cooled VW Engines: 0-89586-225-5
Rebuild Big-Block Chevy Engines: 0-89586-175-5
Rebuild Big-Block Ford Engines: 0-89586-070-8
Rebuild Big-Block Mopar Engines: 1-55788-190-1
Rebuild Ford V-8 Engines: 0-89586-036-8
Rebuild Small-Block Chevy Engines: 1-55788-029-8
Rebuild Small-Block Ford Engines:0-912656-89-1
Rebuild Small-Block Mopar Engines: 0-89586-128-3

RESTORATION, MAINTENANCE, REPAIR

Camaro Owner's Handbook ('67–'81): 1-55788-301-7
Camaro Restoration Handbook ('67–'81): 0-89586-375-8
Classic Car Restorer's Handbook: 1-55788-194-4
Corvette Weekend Projects ('68–'82): 1-55788-218-5
Mustang Restoration Handbook('64 1/2–'70): 0-89586-402-9
Mustang Weekend Projects ('64–'67): 1-55788-230-4
Mustang Weekend Projects 2 ('68–'70): 1-55788-256-8
Tri-Five Chevy Owner's ('55–'57): 1-55788-285-1

GENERAL REFERENCE

Auto Math:1-55788-020-4
Fabulous Funny Cars: 1-55788-069-7
Guide to GM Muscle Cars: 1-55788-003-4
Stock Cars!: 1-55788-308-4

MARINE

Big-Block Chevy Marine Performance: 1-55788-297-5